KV-428-657

Strategic Value Analysis

BISHOP BURTON LRC
WITHDRAWN

ACCESSION No. T048111
CLASS No. 658.4012 Ref

In an increasingly competitive world, we believe it's quality of thinking that will give you the edge – an idea that opens new doors, a technique that solves a problem, or an insight that simply makes sense of it all. The more you know, the smarter and faster you can go.

That's why we work with the best minds in business and finance to bring cutting-edge thinking and best learning practice to a global market.

Under a range of leading imprints, including *Financial Times Prentice Hall*, we create world-class print publications and electronic products bringing our readers knowledge, skills and understanding which can be applied whether studying or at work.

To find out more about our business publications, or tell us about the books you'd like to find, you can visit us at www.business-minds.com

For other Pearson Education publications, visit www.pearsoned-ema.com

Strategic Value Analysis

Organize your company for strategic success

TAKEO YOSHIKAWA, JOHN INNES AND FALCONER MITCHELL

An imprint of Pearson Education

London ▪ New York ▪ Toronto ▪ Sydney ▪ Tokyo ▪ Singapore ▪ Hong Kong ▪ Cape Town
New Delhi ▪ Madrid ▪ Paris ▪ Amsterdam ▪ Munich ▪ Milan ▪ Stockholm

PEARSON EDUCATION LIMITED

Head Office:
Edinburgh Gate
Harlow CM20 2JE
Tel: +44 (0)1279 623623
Fax: +44 (0)1279 431059

London Office:
128 Long Acre
London WC2E 9AN
Tel: +44 (0)20 7447 2000
Fax: +44 (0)20 7240 5771
Website: www.briefingzone.com

First published in Great Britain in 2002

© Pearson Education Limited 2002

The right of Takeo Yoshikawa, John Innes and Falconer Mitchell to
be identified as authors of this work has been asserted by them in
accordance with the Copyright, Designs and Patents Act 1988.

ISBN 0 273 65429 2

British Library Cataloguing in Publication Data
A CIP catalogue record for this book can be obtained from the British Library.

All rights reserved; no part of this publication may be reproduced, stored
in a retrieval system, or transmitted in any form or by any means, electronic,
mechanical, photocopying, recording, or otherwise without either the prior
written permission of the Publishers or a licence permitting restricted copying
in the United Kingdom issued by the Copyright Licensing Agency Ltd,
90 Tottenham Court Road, London W1P 0LP. This book may not be lent,
resold, hired out or otherwise disposed of by way of trade in any form
of binding or cover other than that in which it is published, without the
prior consent of the Publishers.

10 9 8 7 6 5 4 3 2 1

Typeset by Monolith – www.monolith.uk.com
Printed and bound in Great Britain by Ashford Colour Press Ltd, Gosport, Hants.

The Publishers' policy is to use paper manufactured from sustainable forests.

About the authors

Takeo Yoshikawa is Professor of Management Accounting at Yokohama National University, Japan and Visiting Professor at the University of Edinburgh.

John Innes is Professor Emeritus of Management Accounting at the University of Dundee.

Falconer Mitchell is Professor of Management Accounting at the University of Edinburgh.

The three authors have been jointly researching SVA since the late 1980s and some of this research was funded by the Canon Foundation. Most of this research has been either case studies of companies or action research involving actual SVA exercises with Japanese and Scottish organizations.

Contents

List of figures

Acknowledgements

The authors wish to acknowledge the Canon Foundation which awarded a visiting professorship to Takeo Yoshikawa and visiting fellowships to John Innes and Falconer Mitchell during the 1990s and so facilitated the international collaboration necessary to write this book. Takeo Yoshikawa wishes to acknowledge the Grant-in-Aid for Scientific Research by the Ministry of Education, Culture, Sports and Technology (MEXT).

We also wish to thank Marsha Caplan for her skill and perseverance in typing the manuscript and Laurie Donaldson for his editorial advice and encouragement. Finally, we are indebted to all the accountants and managers who have over the last 15 years, co-operated with us in our research projects and so allowed three academics to maintain contact with the 'real' world.

Executive summary

Strategic value analysis (SVA) is a multidisciplinary team approach which can support:

1 the design and development of new products and services;
2 the redesign of existing products and services;
3 the management of overheads;
4 the development of strategy and related performance measures;
5 decision-making.

The design stage is critical for many organizations because over 80 per cent of the total costs for a product or service may be committed (but not incurred) during the design stage.

Strategic value analysis concentrates on the individual functions rather than on the individual parts of a product. For example, the functions of a pen include:

1 make mark;
2 add colour;
3 flow ink;
4 store ink;
5 hold pen;
6 prevent stains.

The basic approach is to take account both of existing costs and of customers' views of such functions so that problem functions can be identified and subsequent solutions generated during the SVA team brainstorming session.

Techniques which aid the SVA process such as target costing, cost tables and Kousuu are discussed. Detailed examples of SVA are given in various case studies of both products and services. Strategic value analysis worksheets illustrate how a structured approach can be used to generate not only significant cost reductions but also features which give a competitive advantage. Strategic value analysis is an approach which takes account of both cost and value to the customer, and helps to develop and implement an organization's strategy.

1

Introduction

Finding ways of improving performance is a central and continual challenge for corporate management. This book is based on a technique called strategic value analysis. This technique's origins lie in value engineering and value analysis which were developed in the West (Creasy, 1973) and used by many leading companies. While in the West the technique remained in the domain of engineering, it was developed to encompass many other disciplines since 1960 in Japan to help meet the challenge. During this period Japanese business emerged from relative obscurity to become international leaders as corporate performance improved dramatically. This was achieved against the background of a country with few natural resources and with a phenomenal currency appreciation which certainly challenged, but did not defy, the country's export-led rise to economic prominence.

Strategic value analysis has played an important role in this success story. It has provided both a structured approach to meeting the challenge of performance improvement and the mechanic for generating the necessary ideas and actions which underlie a successful competitive strategy. Strategic value analysis has also proved ubiquitous as it has developed from a focus on the product/customer interface to extend to most of the key organizational, planning and decision-making areas. It brings together people, information and clarity of objective in a systematic procedure aimed at delivering innovative, constructive and financially rewarding change.

This book contains an explanation of SVA which is designed to facilitate its adoption. It shows how the technique has been used in practice and provides examples and procedural guides to its various areas of application. By way of introduction this chapter outlines the main aspects of SVA and the key attributes which underlie its successful implementation and use.

THE BASICS OF STRATEGIC VALUE ANALYSIS

Strategic value analysis was developed as a tool to improve product functionality and cost effectiveness. Its origins lie in the techniques of value engineering and value analysis which are specialist engineering-based techniques aimed at improving the value offered by products. However, in Japan this technique was extended and improved to become a much broader and more pervasive approach to business improvement.

In essence, SVA is based on viewing the product not in its existing physical manifestation, but as a structured group of related services or functions which it offers to the customer. Thus a chair would be viewed not simply as the wood and fabric of which it is made, but of the comfort, support and decorative presence which it offered to its user. This more abstract conceptualization focuses on the attributes which give the product, and ultimately the organization, its value. It is a market- or output-oriented reflection of the product. Identification of these

BISHOP BURTON COLLEGE
LIBRARY

service functions is the first step in a process of preserving, amending, developing and exploiting the characteristics of the product output which attract customers. It provides a focus based on the market, and one in which the needs of the customer (both functionality and price) can remain paramount. From this perspective the constraints of what currently exists (in terms of the physical product) do not unduly restrict consideration of how the customer can be serviced. A pipeline can be viewed as an alternative to a fleet of ships, a laser for a knife and glue for a nail. Radical design change is therefore facilitated. Enhancing the product through improving or increasing the services which it offers can also be more easily considered from this perspective. Obtaining an SVA of the product results in a description based on its decomposition into an abstract structuring of its service potential. This gives clear visibility to what it is that the organization offers its customers. In turn, this information can be used as a means of identifying why customers buy the product, what they particularly like about it, what gives it value to them, how it might be changed to increase customer attraction and what customers do not rate so highly about it. Indeed, this can be quantitatively expressed by obtaining customer ratings for all the product functions offered. The result is an accessible distribution of key factors underlying the market success (or otherwise) of the product.

However, value and service offered to customers only comes at a cost and without appreciation and incorporation of this aspect of performance SVA would be deficient. Thus, while functionality provides a basis for SVA, it has to be supplemented by a cost analysis which also fits the approach. Product cost has to be analysed by function or service rather than its conventional cost accounting treatment by type of input (material, labour and overhead). Together these characteristics generate a profile of the products which the organization markets. Figure 1.1 summarizes this information.

The information in Figure 1.1 demonstrates how SVA links the internal operation of the organization with its market context. The relative cost column shows how corporate resources are consumed by the functionality delivered in the product supply, while the customer rating column shows how the customers value each of the components of functionality. In this way the product utility underlying demand can be screened against the capabilities and characteristics of its supply. Comparing these two columns, therefore, provides an important guide for both cost management activity and product (re)design. Incongruencies between the columns based, for example, on higher relative costs than the corresponding customer value ratings (e.g. function 2 in Figure 1.1) indicate where economies may be targeted based on market opinion. In addition, product improvement and investment can be similarly guided where the customer ratings show that the relative importance of a function exceeds the commitment of resource provision (function 1 in Figure 1.1 above). Thus the SVA approach provides a signalling system which, in the first instance, directs managerial attention to specific product attributes. However, the strength of SVA is not simply in identifying problems and opportunities. It also provides a set of procedures to follow in order to generate possible solutions in respect to the signal indications. In this lies its great strength; it is a constructive approach designed to identify and promote positive improvement within the organization. Achievement of these end results can be attributed to five powerful aspects of the SVA approach.

Fig. 1.1 Key functional analysis information

Product XXX:

	Cost (£)	Relative cost (%)	Customer rating
Function profile:			
Function 1	£x	10%	15%
Function 2	£y	27%	12%
Function 3	£z	16%	16%
⋮	⋮		⋮
Function n	£n	n%	n%
Total	£a	100%	100%

1 Effective utilization of intellectual capital

The skill, knowledge and expertise of staff represent the most valuable asset of many businesses. Strategic value analysis provides a means of ensuring that the potential benefits of this asset are fully enjoyed. This is achieved by providing a forum where key staff members interact to generate improvements to various aspects of the organization and its products or services (both internal and external). All functional disciplines (e.g. research, production, purchasing, distribution, sales and administration) can contribute suggestions on product or service design modification or, indeed, on more radical change often involving new developments. Moreover, they can all assess how to revise their own special contribution in order to save costs and/or improve customer service. It is, however, the bringing together of staff from these different areas which stimulates the 'brainstorming' sessions and the creativity which can release the true power of the firm's intellectual capital.

2 Clear purpose orientation

Strategic value analysis is so powerful because it brings a clear focus to this activity. Not only is there a clear object for the activity (e.g. particular products or services), but clear targets can also be set for the interdisciplinary teams in terms of cost savings levels and/or product improvements (e.g. weight loss, size reduction or patent applications). Thus the SVA team knows exactly what is expected of it. The establishment of a highly procedural approach also provides a set of steps to be taken in pursuit of these objectives. Although these give a structured framework for the activity, within this there is plenty of scope for creative and innovative thinking. The SVA approach facilitates the generation of good ideas of all types. It does not restrict or constrain but, rather, encourages change and improvement directed to a particular purpose.

3 Value creation

Strategic value analysis can support the twin aspects of competitive strategy which deliver future value – product price and product differentiation. Through the analysis of cost and the identification of cost reduction opportunities SVA can provide a foundation for cost (and therefore price) advantage. For example, it can assist in moves to pursue the achievement of low cost producer in a sector. Moreover, this strategy is guided in SVA by market sentiment to ensure that the customer service implications of changes in resource consumption do not destroy demand. Through the focus on product attributes SVA ensures that the organization remains sensitive to the functionality demanded by customers now and in future. It thus promotes and guides the change and flexibility required to maintain a successful differentiation policy.

4 Ordered structuring issues

Strategic value analysis provides a means of addressing issues and challenges in a structured way. It provides a systematic approach to the process of analysing and developing solutions. At the core of this ordering of the 'situation' lies the construction of a set of means/end relationships relating to the object (product, service, decision, plan, etc.). This is constructed in diagrammatic form which facilitates description and analysis of the issue.

For example, the technique might be used to reconcile and integrate different views on the purpose of a service function. If asked about the purpose of the management accounting function, a set of apparently disparate opinions, listed below, might well be gathered from a group of managers.

1 To provide financial information for decision makers.

2 To initiate and justify investment decisions.
3 To help meet corporate objectives.
4 To track performance.
5 To control financial performance in accordance with plans.
6 To help operational managers select the most appropriate work plans.
7 To assess the performance of people and parts of the organization.
8 To identify where plans are not being met.
9 To translate plans into financial terms.
10 To analyse cost behaviour.
11 To reconcile performance with the financial accounts.
12 To produce the monthly accounting package.

When these are restructured into a means/end structure they transform the confusing listing of multiple purposes into a much clearer formulation of the nature of, and contribution made by, management accounting (see Figure 1.2). This presentation shows how the multiple contributions of management accounting are achieved and how they are related to each other. It thus provides an appropriate starting point for assessing, and possibly redesigning, the system.

Fig. 1.2 **Diagrammatic analysis of the contribution of management accounting**

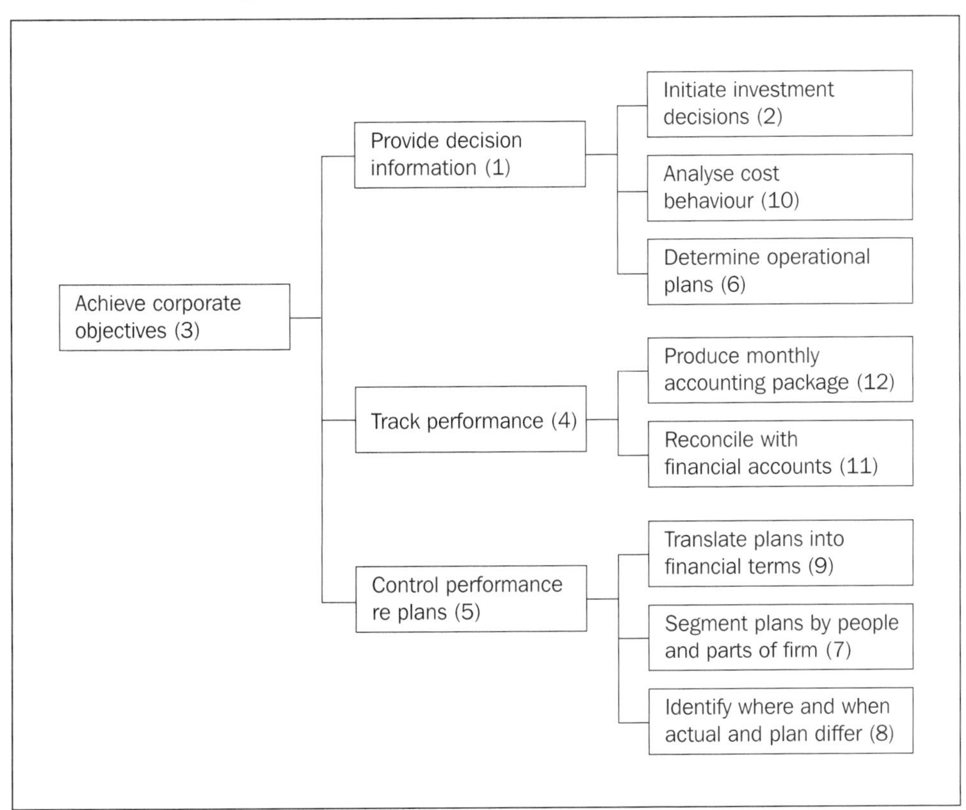

5 Multiple applications

While SVA was originally developed as a product-oriented technique, it has now been extended to provide an approach which can be applied in a variety of settings.

This book outlines its use not only in designing or redesigning products (Chapter 2), but also in improving both internal services which constitute the overhead of the organizations and the external services sold to customers (Chapter 8) and the analysis of decisions and the design of strategic performance measurement systems (Chapter 9). Chapter 3 provides detailed information which will allow either a product or service organization to implement SVA and takes you step-by-step through ten SVA worksheets. Chapters 4, 5 and 6 give details of three techniques which aid the SVA process, namely, target costing, cost tables and Kousuu. Chapter 7 is an SVA case study illustrating the use of SVA by one organization. The versatility of the approach is one of the underlying strengths which can be exploited by the experienced SVA user.

OVERVIEW

For all the above reasons SVA is a technique which can have a powerful, enduring and highly positive effect on organizational performance. To a large extent the technique can be programmed as a series of easily explained steps which have to be followed. However, its success depends on the staff who undertake to run SVA exercises and their ability to deliver useful results does improve with experience of the technique. Indeed, staff time is the greatest investment required to use SVA but when adopted and operated properly it will repay that investment many times.

2

SVA for products

DESIGN AND DEVELOPMENT OF PRODUCTS

Strategic value analysis is a useful approach both during the design and development of new products and the redesign of existing products. The design stage is critical because, for many organizations, over 80 per cent of the total costs for a product or service are committed during the design stage. Most costs are not incurred usually until during the production stage but the design decisions commit the organization to incurring these costs. In our experience design engineers have a good understanding of direct material and direct labour costs but often do not understand what drives overhead costs. If management accountants are not involved as members of your design teams, you should be asking your designers why this is the case. Similarly, if management accountants are not involved in your design process, you should be asking them why, if 80 per cent of the costs are committed during the design stage, they are not spending 80 per cent of their time on the design process.

Strategic value analysis concentrates on the individual functions rather than on the individual parts of a product. For example, the parts of a pen include the barrel, ink cartridge, tip and top. However, the functions of a pen include:

1 make mark
2 add colour
3 flow ink
4 store ink
5 hold pen
6 prevent stains.

By concentrating on the functions of a product rather than on its parts, you take a much more general view of the product. This is very helpful during both the design of new products and the redesign of existing products. For example, if you are redesigning an existing product and you concentrate on the parts of that product, almost certainly the redesigned product will be very similar to the existing product. In contrast, by thinking in terms of the functions of the existing product, the redesigned product may look very different from the existing product. Indeed, the SVA process can lead to the development of new patents and, even, new products.

OBJECTIVE – PROFIT IMPROVEMENT

The objective of SVA for products is to increase profits and not just reduce costs. Strategic value analysis can help to identify functions of a product where customers wish more to be spent, leading to increased profits for the organization.

In addition SVA may identify new functions for a product which will make the product more attractive to customers and, again, lead to improved profits.

Sometimes SVA will concentrate on a problem area of a product. For example, the bulk and weight of a component to be included in a motor car may be a problem. The objective of the SVA exercise may be to reduce the bulk and weight of this particular product by 25 per cent. However, although this is the objective of this particular exercise, again the underlying objective is to make this product more attractive to the customer (in this case the car manufacturer) so that a higher price may be charged, leading to higher profits.

STEPS IN SVA

A very important aspect of SVA is that it is a team activity. Strategic value analysis gives a structured approach to achieving a specific objective to which all team members can contribute. In Japan there are usually competing SVA teams designing the same product. However, when a Scottish organization ran an SVA exercise, two teams were involved but they co-operated with each other because one team had an intimate knowledge of the product and the other team knew nothing about the product. Interestingly, the best suggestions came from the team with no knowledge of the product. The steps in any SVA exercise are as follows:

1 *Select products for SVA.* In Japan, organizations have a department with employees who select the products for the SVA exercises. For new products, the SVA can be integrated into the design process. For existing products or components a critical part of SVA is the selection process. The criteria for selection include:

   (a) very complex;

   or (b) very heavy;

   or (c) very bulky;

   or (d) customer requests;

   coupled with a relatively high cost.

2 *Decide specific objective.* The criteria for selecting the product for an SVA exercise will help to determine the specific objective, such as weight reduction by 20 per cent coupled with cost reduction of 40 per cent without reducing the quality of the product. Very often SVA is combined with target cost management (see Chapter 4) where the cost is determined from an anticipated future selling price. For example, if the anticipated selling price is £100 per unit and the profit margin is £10 per unit, then the target cost is £90 per unit. However, the existing cost per unit is £150 and the cost per unit therefore needs to be reduced from £150 to £90, i.e. 40 per cent cost reduction for SVA exercise.

3 *Plan schedule.* It is important to determine a schedule for each SVA exercise. The actual schedule depends on the complexity of the exercise. The time taken can vary from one week to several months but most SVA exercises can be completed in under a month.

4 *Select team.* The number in the SVA team (or teams) depends on the exercise. However, a team leader will need to be identified who can take the initiative for running the SVA exercise and achieving its objective. The other team members will have different skills such as design, engineering, management accounting, production, purchasing and sales. The team will usually consist of four to six employees who are seconded temporarily from their normal jobs. A full-time employee from the SVA department will usually provide advice to the team leader and the team.

5 *Collect information.* At the beginning of the SVA process it is important to collect the necessary information such as the following:

(a) existing design;

(b) existing specifications for product;

(c) details of production process;

(d) material, labour and overhead cost data;

(e) scrap data;

(f) marketing data about the product.

6 *Decide product functions.* It is important that the team thinks about the functions rather than the parts of the product. This involves a brainstorming process among the team members. The functions are expressed in terms of a verb and a noun such as 'make mark' for the main function of a pen. It is useful to relate the different functions together and you can do this by asking the question 'How?'. For example, how do you make a mark? The answer is 'add colour', and if you ask how do you add colour the answer is 'flow ink'. All the different functions can be grouped together in a functional family tree which will be illustrated in the case study which follows these 13 detailed steps.

7 *Calculate cost of each function* (*see* Yoshikawa et al., 1989). This is a traditional costing exercise to calculate the cost of each function for a pen, such as 'add colour' and 'flow ink'. The existing costs for a pen are:

	Pence
Add colour	4
Hold pen	6
Flow ink	6
Store ink	10
Prevent stains	6
Total	32

8 *Determine customers' values for each function.* Just as target costing is a different form of costing by starting from the market, so a critical element for SVA is bringing in the customers' views. Perhaps the best way to do this is by market research asking customers to rate all the functions so that they total 100 per cent – for example for a pen:

	%
Add colour	20
Hold pen	10
Flow ink	30
Store ink	10
Prevent stains	30
Total	100

9 *Assign target cost to each function.* The customer derived percentages (under step 8 above) can then be used to assign the target cost to each function. For example, if the target cost set for a pen is 20 pence, then this target cost would be assigned to each function as follows:

	Pence	
Add colour	4	i.e. 20% × 20 pence
Hold pen	2	i.e. 10% × 20 pence
Flow ink	6	i.e. 30% × 20 pence
Store ink	2	i.e. 10% × 20 pence
Prevent stains	6	i.e. 30% × 20 pence
Total	20	

10 *Determine problem functions.* The existing costs of each function (from step 7 above) can be compared with the assigned target cost (from step 9 above) to determine the problem functions:

	Existing costs	*Target cost assigned by customers' weightings*
	Pence	*Pence*
Add colour	4	4
Hold pen	6	2
Flow ink	6	6
Store ink	10	2
Prevent stains	6	6
Total	32	20

The above suggests that hold pen and store ink are the two problem functions where the existing cost is much higher than the target cost assigned by the customers' weightings. This type of market-based approach is a critical feature of SVA.

11 *Brainstorming possible solutions.* The most important step in SVA is the brainstorming among the team members where all make suggestions for improvement. No suggestion is rejected at the initial stage because what at first seems to be a silly suggestion may turn out to be the best solution in the final analysis. Each suggestion needs to be costed and this is where cost tables (see Chapter 5) are very helpful. Similarly Kousuu (see Chapter 6) can be useful during this brainstorming processing. The suggestions include deleting functions, combining functions, using new materials or using new production processes. In the pen example, the SVA team would concentrate on suggesting solutions to the 'hold pen' and 'store ink' problems.

12 *Decide solution and present results.* The SVA team chooses the best solution from all the suggestions to achieve the required objectives and target cost. All team members present to two or three top managers in the organization the findings, the suggested changes and the expected benefits. This climax to the process stimulates competitiveness since no team member wants to lose face by failing to achieve the objectives set for this SVA exercise. These presentations are formal and each team also prepares a report about its own SVA exercise.

13 *Implement solution and audit results.* It is obviously important to implement the chosen solution and ensure that the predicted results are actually achieved in practice. To prevent overoptimistic assessments of the SVA results, the team members know that within a year an auditor will report to top management on the actual effects of the changes implemented because of the SVA exercise.

Case study

A manufacturing organization has used SVA for many years for a variety of purposes, including the redesign of existing products. One example will be given in this case study, namely, the redesign of a staple remover. Figure 2.1 explains that the existing staple remover is too heavy and difficult to handle and the existing unit cost is 135 yen but the target cost is only 85 yen (steps 1 and 2 above). Figure 2.1 also gives information (step 5) about the staple remover including design, cost constraints on design and requests from customers.

The sixth step is to decide the functions of the staple remover. Figure 2.2 shows the basic functions for the staple remover in the form of a functional family tree.

The seventh step is to calculate the cost of each function and Figure 2.3 summarizes these costs.

The eighth step is to determine the customers' evaluations of all the functions, and these are expressed in Figure 2.4 in terms of the 'expected cost' which is the customers' ratings of the functions expressed in terms of the target cost of 85 yen (ninth step). Figure 2.4 also compares the customers' expected cost with the actual cost to determine the most significant problems for the tenth step. The most important problem is 'hold staple' with customers' expected cost of 22 yen against actual cost of 50 yen.

Fig. 2.1 Information for strategic value analysis

☐ **General**

The purpose of the product, a staple remover, is to remove a staple without tearing the paper.

☐ **Existing parts of the product**

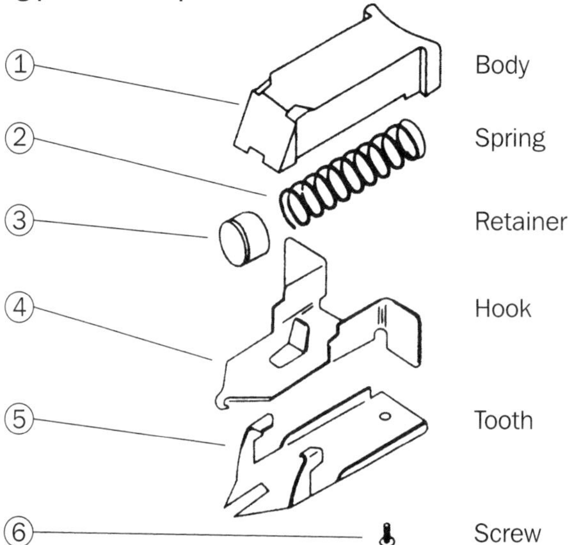

☐ **Costs of the parts**

No.	Part name	Weight (grams)	Cost (Yen) Material	Conversion	Total
1	Body	5.0	1.0	19.0	20.0
2	Spring	1.0	1.0	7.0	8.0
3	Retainer	0.5	0.2	4.8	5.0
4	Hook	11.5	3.0	42.0	45.0
5	Tooth	11.5	3.0	47.0	50.0
6	Screw	0.5	0.1	1.9	2.0
7	Assembly	—	—	5.0	5.0
	Total	30.0	8.3	126.7	135.0

☐ **Requests from customers**

- The staple remover should be lighter and easier to handle.
- The staples should not be scattered about.

☐ **Constraints on design**

- The size of staples to be removed is 9mm x 4mm.
- The product should be used in one hand.
- Elementary-age schoolchildren should be able to use the product.
- The staple remover should last for more than 100,000 uses.
- The product should not rust for at least five years.
- The target cost is Yen 85.

Source: Yoshikawa et al. (1996).

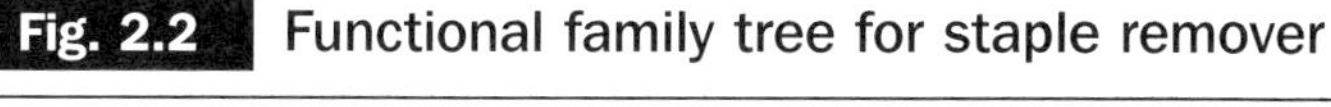

Fig. 2.2 Functional family tree for staple remover

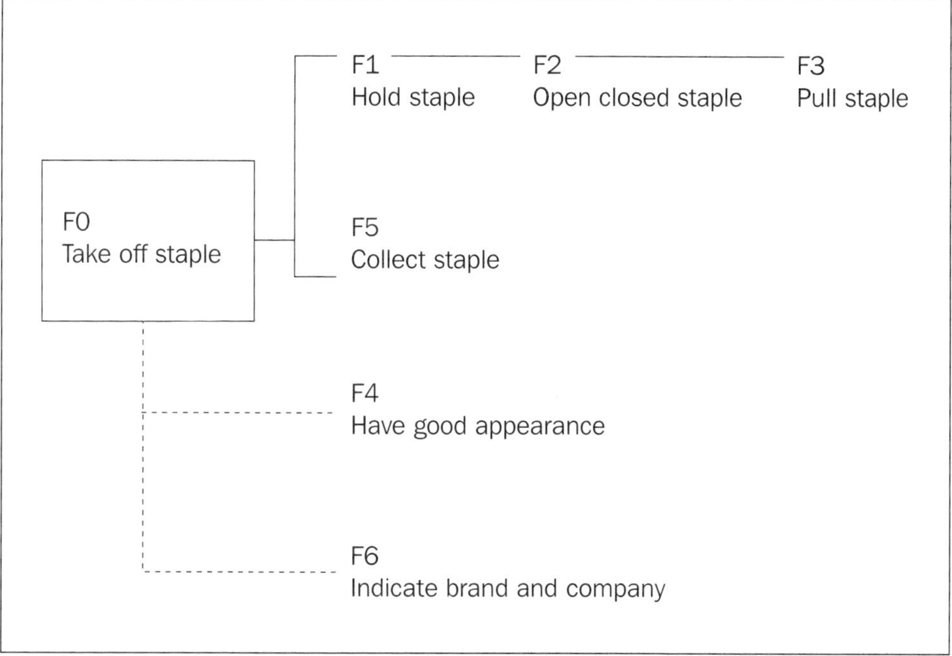

Fig. 2.3 Actual cost of each function

	Yen
Hold staple	50
Open closed staple	26
Pull staple	26
Have good appearance	16
Prevent staples from scattering	16
Indicate brand and company	1
Total Cost	135

The eleventh step is the brainstorming where the SVA team decided to combine a number of functions which reduced the number of parts in the staple remover (see Figure 2.5). In the twelfth step of deciding the solution and presenting the results, the spring, retainer and screw were eliminated and the body of the staple remover, hook and tooth were combined together. The redesigned staple remover is shown in Figure 2.5 with a reduced cost per unit of 43 yen against the existing cost of 135 yen and the target cost of 85 yen. Figure 2.5 also calculates an annual net saving of approximately 20 million yen on an annual production volume of 240 000 staple removers. This particular solution was implemented and the redesigned staple remover is now on sale. About six months later the results from the redesigned staple remover were audited and the net saving was in fact greater than the 20 million yen forecast by the SVA team.

Fig. 2.4 Evaluation of functions

No.	*Function*	*Actual Cost (Ca) (yen)*	*Expected Cost (Ce) (yen)*	*Ce/Ca*	*Rank*
1	Hold staple	49.6	22.0	0.4	1
2	Open closed staple	26.2	21.0	0.8	3
3	Pull staple	26.2	21.0	0.8	3
4	Have good appearance	16.0	8.0	0.5	2
5	Prevent staples from scattering	16.0	12.0	0.8	5
6	Indicate brand and company	1.0	1.0	1.0	6
Total		135.0	85.0[a]	0.6	

Note: [a]The expected cost of 85 yen is also known as the target cost.

The redesigned staple remover is also lighter, easier to use and collects the staples. However, the SVA team recommended that the function of collecting the staples could be improved and could be the objective of another SVA exercise. Nevertheless, a valuable new function of collecting the staples has also been added to the product and this was an additional function requested by customers.

Fig. 2.5 Suggested staple remover

Selected alternative: one of the tweezer types

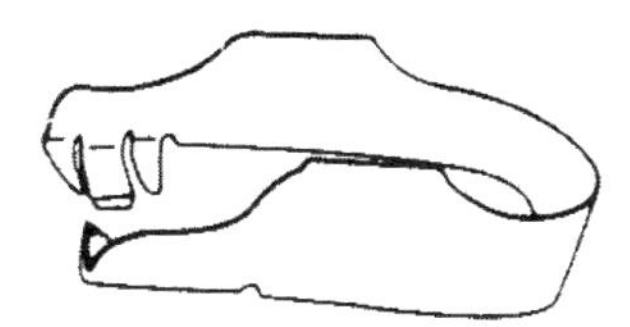

1 *Current cost*: Yen 135

2 *Target cost*: Yen 85

3 *Cost of selected alternative*: Yen 43.1/unit

Material/unit: Yen 19.1

Conversion cost/unit: 24.0

4 *Other costs*: Yen 4,344,310

Design: Yen 4,000 × 20 manhours = 80,000

Mold: = 4,100,000

Yen 600,000 × 1 = 600,000

Yen 700,000 × 5 = 3,500,000

Trial production: Yen 431 × 10 = 4,310

Test: Yen 4,000 × 40 manhours = 160,000

5 *Savings*: (1) – (3) = Yen 91.9/unit

6 *Saving rate*: (5) / (1) = 68.1%

7 *Achieved rate of target cost*: (2) / (3) = 199%

8 *Production volume*: 20,000 units × 12 = 240,000 units

9 *Annual saving*: (5) × (8) = Yen 91.9 × 240,000 = Yen 22,056,000

10 *Annual net saving*: (9) – (4) X 0.5 (2-year depreciation)

Annual saving: Yen 22,056,000

Less: Other costs: 2,172,155

Yen 19,883,845

Source: Yoshikawa et al. (1996).

3

SVA worksheets

INTRODUCTION

To run SVA exercises, the teams find it helpful to have a series of worksheets which help to give a structured approach to the exercise. Having chosen the object of analysis such as a product or service and determined the target to be achieved by the SVA activity, it is necessary to plan the schedule for implementing the SVA exercise and to set up the SVA team (or teams).

It is during the next implementation stage that the ten SVA worksheets are used for the following steps in the process.

1. Collect information about characteristics (volume, cost ratio to selling price, number of parts, etc.) of the object of the SVA.
2. Collect further information (use, marketing, design, purchasing, manufacturing, etc.) about the object of the SVA.
3. Decide functions of the object of the SVA.
4. Draw a functional family tree.
5. Calculate cost of each function.
6. Evaluate functions including customers' values for each function.
7. Generate alternative suggestions for improvements to object of SVA.
8. Draw revised functional family tree following changes to the object of the SVA.
9. Summarize economic and engineering information about changes to the object of the SVA.
10. Give final recommendation including profit improvement.

In addition to providing a structured approach for SVA exercises, these ten worksheets also give a useful record of SVA exercises. The use of these ten worksheets will be illustrated in the following case study.

Case study

YIM Ltd was founded in 1900 and soon became a leading pen manufacturing company. About 70 years ago YIM developed a new type of fountain pen which became a market leader in many countries. However, eventually the sales of YIM Ltd began to fall because customers disliked certain characteristics of fountain pens, namely:

1. need bottle of ink to refill pen;
2. pen ran out of ink very often;
3. ink stained customers' hands;
4. fountain pen nib was easily damaged.

BISHOP BURTON COLLEGE
LIBRARY

The next development in YIM Ltd was that its research and development department developed an ink cartridge with much thicker ink and this led to the propelling ballpoint pen which sold very well because of its:

1 very smooth writing;
2 very long-lasting pen point;
3 luxurious appearance making it a gift for special occasions.

However, eventually the sales of this relatively expensive propelling ballpoint pen began to decline because many customers already had more than one such pen and were now more interested in a disposable pen with good environmental features and a lower price. YIM Ltd therefore decided to conduct an SVA exercise on their existing propelling ballpoint pen.

Ten worksheets

The ten worksheets in this case study are modified versions of those suggested by the Sanno Institute of Management VM Centre (1995). The first worksheet summarizes the characteristics and reasons for selecting the pen for the SVA exercise, including the number of pens manufactured, the cost/sales ratio, number of parts in pen, requirements by customers, possibility of improvements and reasons for selection of pen for SVA. The second worksheet includes further information about the pen including its application (i.e. use) marketing, design, purchasing, manufacturing existing cost and target cost.

The third worksheet is where the parts of the pen are shown but, much more importantly, where the team lists the various functions of the pen such as flow ink, guide nib, store ink and put colour (the four basic functions) and also the secondary functions of prevent stains and prevent loss. The restriction column on the third worksheet would include any restrictions on the functions such as government regulations or particular specifications from customers.

Having determined the functions of the pen, the team draws a functional family tree on the fourth worksheet. The main or primary function is make mark and the other basic functions are shown by solid lines in the diagram on the fourth worksheet and the secondary functions are denoted by dotted lines. The logic of this functional family tree can be checked by asking the question 'how' going from the left to the right of the diagram and by asking the question 'why' going from the right to the left of the diagram. For example, how do you 'make a mark'? The answer is to 'put colour'. And how do you 'put colour'? The answer is to 'flow ink'. Similarly in reverse, why do you 'flow ink'? The answer is to 'put colour'. And why do you 'put colour'? The answer is to 'make mark'.

The fifth worksheet gives the cost calculation for each function showing the analysis of the existing cost of £4 per pen by functions such as 'flow ink' £1.03 and 'pull in/out nib' £1.54.

The market aspect is formally incorporated into the SVA process in the sixth worksheet including the customers' assessment of the relative value of each function in relation to the target cost of £2.00. These customer assessments are shown in the expected column (Ce) in the sixth worksheet. For example, the customers assessed the function of 'flow ink' as being 25 per cent of the target cost of £2.00 giving 50 pence. This expected cost of 50 pence for 'flow ink' can then be compared with the present cost of flow ink (Cp) of £1.03 from the fifth worksheet. The ratio of:

$$\frac{\text{Customers' expected cost of function (Ce)}}{\text{Present cost of function (Cp)}}$$

can be calculated for each function. The reason for calculating Ce/Cp is to identify potential problem functions where the ratio is less than 1. For example, Ce/Cp for the flow ink function is

$$\frac{£0.50}{£1.03} \text{ i.e. } 0.49$$

However, such problem functions may be low cost functions and therefore it is also useful to calculate Cp-Ce for each function to determine the monetary difference between the present cost of each function and the customers' expected cost of each function. For example, for the 'flow ink' function Cp-Ce was £1.03 less 50 pence, i.e. 53 pence. The sixth worksheet identifies the problem functions in terms of priority with 'flow ink' and 'store ink' being problem functions but the most serious problem function being 'pull in/out nib'.

Having identified the problem functions, the team can list their ideas for improving the pen and achieving the target cost on the seventh worksheet. Usually each suggestion will require a worksheet of its own but for illustrative purposes three suggestions are included on the seventh worksheet. The first suggestion is to use recycled material, and economic and engineering evaluations of each suggestion are included on the seventh worksheet. The team voted for the third suggestion, namely, to make the barrel of recycled material, design a new cap made of recycled material (to combine the prevent stains, attach clip and prevent loss functions) and eliminate the 'pull in/out nib' function. These changes will mean a significant cost reduction and also give a competitive advantage from the environmental viewpoint.

The revised functional family tree is then included on the eighth worksheet showing the simplification of the propelling ballpoint pen to a ballpoint pen by eliminating the 'pull in/out nib' function and combining the prevent stains, attach clip and prevent loss functions.

More detailed information about the chosen alternative is given on the ninth worksheet. For example, subcontractors were found who could manufacture certain parts for the redesigned pen and the design engineers agreed to design a new cap and to make the barrel of the pen from recycled paper. The ninth worksheet also confirms that subcontracting certain parts would lead to a cost reduction and the management accountants confirmed that the target cost of £2 for the redesigned pen had been achieved. Furthermore, market research revealed that the use of recyclable material will double the potential market from 120 000 to 240 000 pens per year.

The final and tenth worksheet summarizes the cost savings of the SVA exercise in the first year although the reduction in the selling price from its existing level to the new lower level would also be taken into account. In this pen example, the cost savings are calculated on the basis of the projected sales of 240 000 pens. The estimated incremental overhead costs are £150 000 and the direct cost savings per unit are £2 for the 240 000 pens, giving net cost savings for the new pen of £330 000 in the first year. With the incremental overhead costs often being one-off in the first year only, the net cost savings should be greater in future years. The tenth worksheet is also a reference sheet for the future and will give a cross-reference to more detailed information and a contact name.

The worksheets are not only very helpful for the team during the SVA experience but also a record for future reference of what has been achieved. This exercise achieved its objectives of:

1 designing a cheaper pen with a lower price;

2 a disposable pen with good environmental features.

However, the main aim of this chapter is to illustrate the use of the ten SVA worksheets which help to provide a useful structure for any SVA exercise.

SVA worksheet 1

CHARACTERISTICS AND REASONS FOR THE SELECTION OF PROJECT FOR SVA
Propelling Ballpoint Pen

1. Number of units (volume) manufacturing:

   120 000 pens manufactured each year

2. Cost ratio (costs/sales)

   £4 cost of pen is relatively high so price is also relatively high

3. Number of parts

   10 parts in pen

4. Requirements by customers

   a) lower price
   b) disposable pen with good environmental features

5. Possibility for improvements.

   Use different materials

6. Reasons for the selection of project

   a) declining sales
   b) declining profits
   c) relatively high cost

TEAM NAME YIM P TEAM MEMBERS T.Y/J.I/F.M/C.P. DATE 1 Sep. 2001

Source: Sanno Institute of Management VM Centre (1995).

SVA worksheet 2

GATHER INFORMATION

PROJECT FOR SVA Propelling Ballpoint Pen QTY./YEAR 120 000 units/year

Information about

1. Application (making use of)

   Use for writing

2. Marketing (selling)

   Most sales are through stationery shops

3. Design

   Existing design is rather old-fashioned

4. Purchasing

   Some parts are purchased from domestic suppliers

5. Manufacturing

   Most parts are manufactured in-house

6. Existing cost

   £4 per pen

7. Target cost

   £2 per pen

TEAM NAME YIM P TEAM MEMBERS T.Y/J.I/F.M/C.P. DATE 1 Sep. 2001

Source: Sanno Institute of Management VM Centre (1995).

SVA worksheet 3

FUNCTIONAL DEFINITION
PROJECT FOR SVA Propelling Ballpoint Pen

No.	Part or assembly		Function			Classification	
			Verb	Noun	Restriction	Basic function	Secondary function
1	Nib		Flow	Ink	No	x	
2	Barrel		Guide	Nib	No	x	
3	Cartridge		Store	Ink	No	x	
4	Top		Store	Ink	No	x	
5	Ink		Put	Colour	No	x	
6	Cap		Pull in/out	Nib	No		x
7	Spring		Pull in/out	Nib	No		x
8	Stopper		Pull in/out	Nib	No		x
9	Nob		Pull in/out	Nib	No		x
10	Clip		Prevent	Loss	No		x

TEAM NAME YIM P TEAM MEMBERS T.Y/J.I/F.M/C.P. DATE 1 Sep. 2001

Source: Sanno Institute of Management VM Centre (1995).

SVA worksheet 4

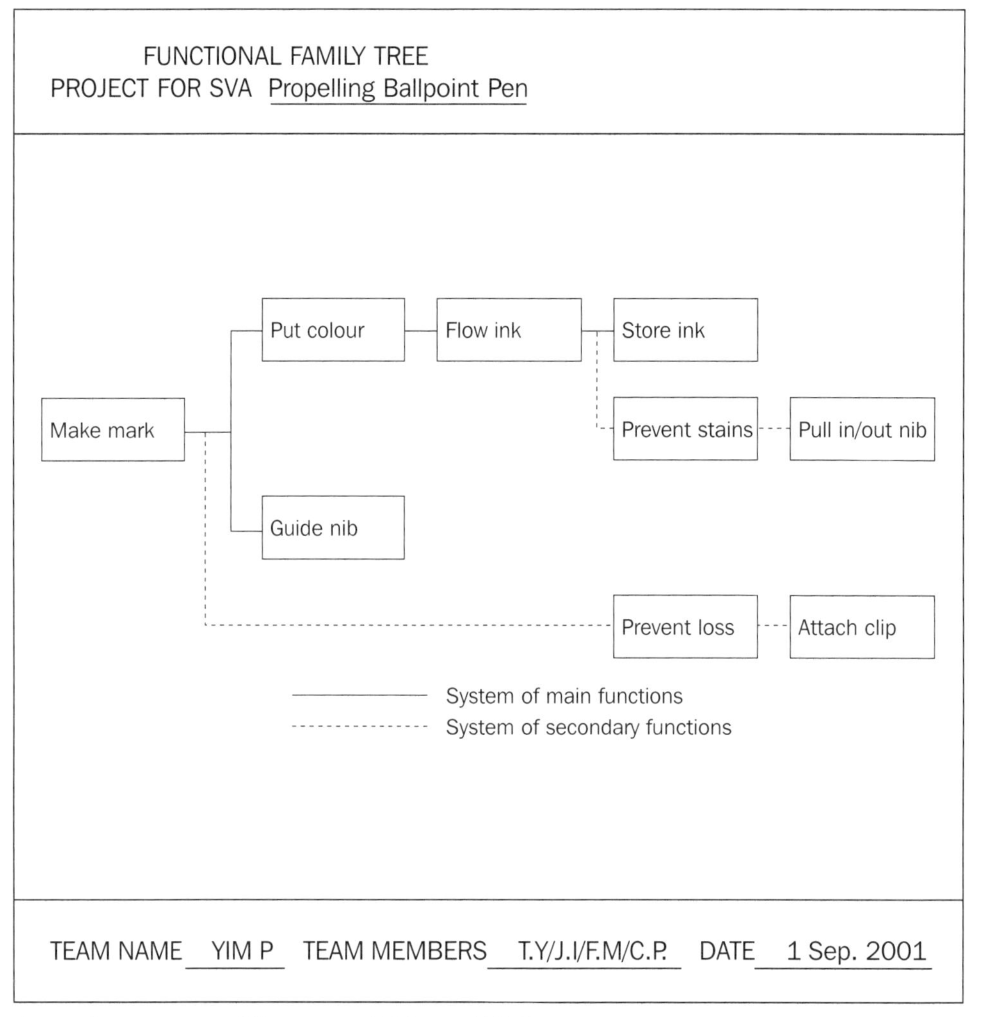

Source: Sanno Institute of Management VM Centre (1995).

SVA worksheet 5

COST ANALYSES OF EACH FUNCTION
PROJECT FOR SVA Propelling Ballpoint Pen

Parts or assemblies \ Function/ field of	Cost	Flow ink	Guide nib	Store ink	Put colour	Pull in/ out nib	Prevent loss and attach clip	
Nib		1.03						
Barrel			0.39					
Cartridge				0.44				
Top				0.20				
Ink					0.19			
Cap						0.74		
Spring						0.40		
Stopper						0.20		
Nob						0.20		
Clip							0.21	
Total	£4.00	£1.03	£0.39	£0.64	£0.19	£1.54	£0.21	

TEAM NAME YIM P TEAM MEMBERS T.Y/J.I/F.M/C.P. DATE 1 Sep. 2001

Source: Sanno Institute of Management VM Centre (1995).

SVA worksheet 6

FUNCTIONAL EVALUATION
PROJECT FOR SVA Propelling Ballpoint Pen

No.	Function/field of function	Expected cost of function (Ce) £	Present cost of function (Cp) £	Ce/Cp	Cp-Ce £	Priority	Notes or coments
1	Flow ink	0.50	1.03	0.49	0.53		Problem function
2	Guide nib	0.56	0.39	1.44	0.17		
3	Store ink	0.30	0.64	0.47	0.34		Problem function
4	Put colour	0.12	0.19	0.63	0.07		
5	Prevent stains and pull in/out nib	0.40	1.54	0.26	1.14	1	Main problem function
6	Prevent loss	0.08	0.18	0.44	0.10		
7	Attach clip	0.04	0.03	1.33	0.01		
Total		2.00	4.00	0.50	2.00		

TEAM NAME YIM P TEAM MEMBERS T.Y/J.I/F.M/C.P. DATE 1 Sep. 2001

Source: Sanno Institute of Management VM Centre (1995).

SVA worksheet 7

<table>
<tr><td colspan="5">FUNCTIONAL DEVELOPMENT
PROJECT FOR SVA Propelling Ballpoint Pen</td></tr>
<tr><td rowspan="2">No.</td><td rowspan="2">Creative idea(s) and development</td><td colspan="3">Evaluation</td></tr>
<tr><td>Economic</td><td>Engineering</td><td>Vote</td></tr>
<tr><td>(1)</td><td>Use different materials, i.e. use recycled material – for example, barrel could be made of recycled paper with clip made of wood.</td><td>Recycled paper is relatively expensive.</td><td>Manufacturing cycle time may be longer.</td><td>NO</td></tr>
<tr><td>(2)</td><td>(a) All parts made of plastic except nib.
(b) Design new cap using plastic.
(c) Eliminate pull in/out function.</td><td>Cost reduction but no competitive advantage.</td><td>Subcontractor could manufacture all parts.</td><td>NO</td></tr>
<tr><td>(3)</td><td>(a) Make barrel of recycled paper.
(b) Design new cap made of recycled paper to combine prevent stains, attach clip and prevent loss functions.
(c) Eliminate pull in/out function.</td><td>Cost reduction and competitive advantage (environmental).</td><td>Manufacturing cycle time may be longer.</td><td>YES</td></tr>
<tr><td colspan="5">TEAM NAME YIM P TEAM MEMBERS T.Y/J.I/F.M/C.P. DATE 1 Sep. 2001</td></tr>
</table>

Source: Sanno Institute of Management VM Centre (1995).

SVA worksheet 8

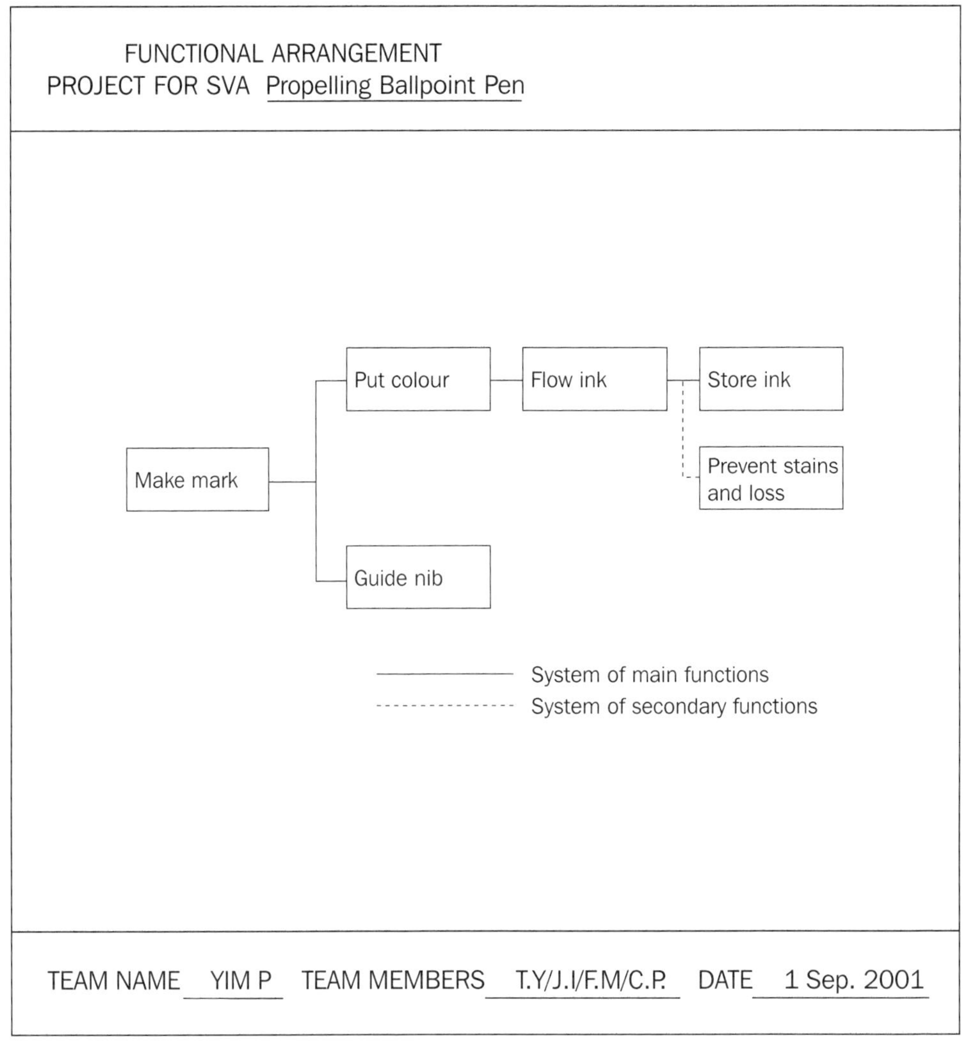

Source: Sanno Institute of Management VM Centre (1995).

SVA worksheet 9

SUBSTANTIVE INVESTIGATION
PROJECT FOR SVA Propelling Ballpoint Pen
IDEA TO DEVELOP No 3

Source of information	Information received	Action taken
Engineering information	1. Subcontractors can manufacture certain parts. 2. Design department can design new cap. 3. No technical problems in making barrel of recycled paper.	1. Subcontractors found. 2. Design engineers agree to design new cap. 3. Commitment to make barrel of recycled paper.
Economic information	1. Cost reduction by subcontracting certain parts. 2. Pen using recyclable material will double the potential market to 240 000 pens per year. 3. Achieved target cost – new pen will cost £2.	1. Completed price negotiations with subcontractors. 2. Completed market research. 3. Accountants agreed detailed costs for new pen.

TEAM NAME YIM P TEAM MEMBERS T.Y/J.I/F.M/C.P. DATE 1 Sep. 2001

Source: Sanno Institute of Management VM Centre (1995).

SVA worksheet 10

SVA RECOMENDATION FORM

		Proposal No. 3
Project Propelling Ballpoint Pen		Company & plant Edin
Specification or Pen Product No. 12345		QTY./Year or Contract period 240 000 units/year

Incremental (estimated) overhead costs	
Item of expenditure	Total £000
Design	100
Trial manufacture	20
Test	20
Other	10
Total	150

	Direct materials £	Direct labour £	Direct other £	Direct cost per unit £
Present cost	2.50	1.25	0.25	4.00
Estimated cost	1.25	0.65	0.10	2.00

(1) Savings of Direct cost/unit	£2
(2) Total Units	240 000
(3) Total Savings of D.C.(1×2)	£480 000
(4) Incremental Overhead Costs	£150 000
(5) Total Net Cost Savings (3-4) in first year	£330 000

TEAM NAME: YIM P TEAM MEMBERS: T.Y / J.I / F.M / C.P.		TEAM No. 1
FURTHER DETAILS: SEE PAGE XXX	CONTACT TEL. XXX	DATE 1 Sep. 2001

Source: Sanno Institute of Management VM Centre (1995).

4

Target costing

BISHOP BURTON COLLEGE
LIBRARY

DEFINITION

An important element of SVA is bringing a market perspective into the cost system and one way to do this is to use target costing. A target cost is defined as:

Target selling price less desired profit

Ideally a target cost for a new product (or service) is set when that product is first being discussed, i.e. before the development and design stages of that product's life cycle. Similarly, a target cost for a redesigned product (or service) is set before the existing product is redesigned.

Traditional costing begins with internal unit costs such as material, labour and overhead, and adds these together to calculate a unit cost. Traditional costing ignores the market completely. In contrast, target costing begins with a future unit price (set by the market) and works back to a target cost. This is useful for SVA where a target cost for the new or redesigned product is set at the very beginning of the process.

The target cost idea is relatively simple although, of course, it is a completely different way of calculating a unit cost from that traditionally used by most organizations. Nevertheless, it is important to be clear about which costs are actually included in the target cost. Most organizations which use target costing include the following costs:

- direct materials
- purchased parts
- labour and processing
- overheads
- depreciation
- development
- trial production
- logistics.

However, a few organizations include only variable costs (such as direct materials, purchased parts, labour and processing) in their target cost but, again, the critical point is that everyone involved knows exactly which costs are included in the target cost.

Stepped and straight line approach

There are also different types of target costs. Figure 4.1 shows the stepped approach and the straight line approach to target costing.

Fig. 4.1 Setting a target cost

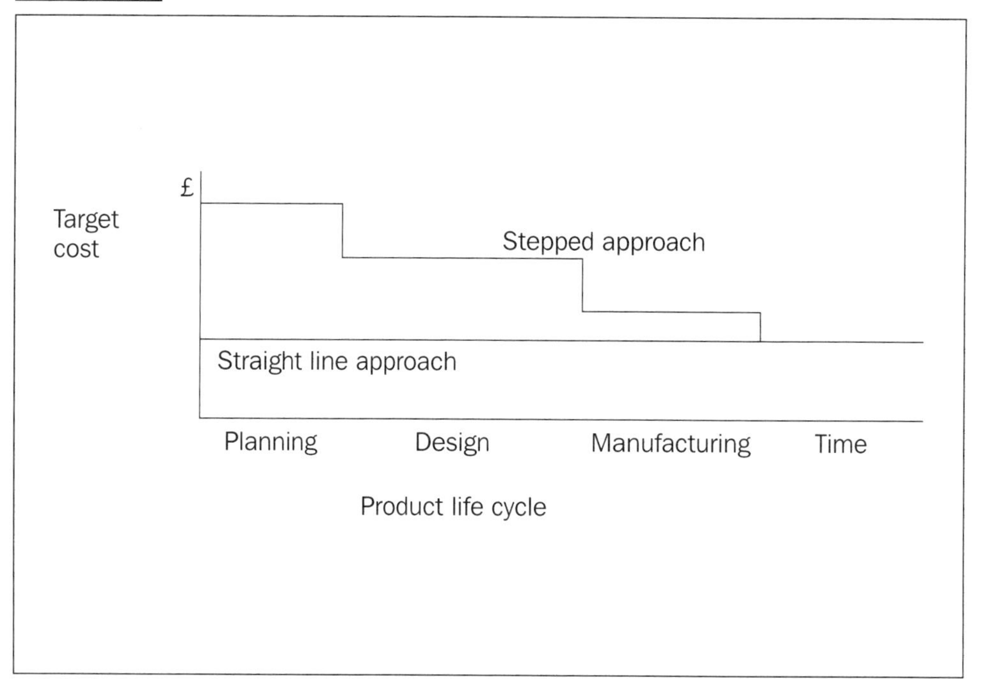

With the stepped approach to target costing, a different target cost is set for each stage of the product life cycle (such as planning, design and manufacturing); whereas, with the straight line approach one target cost is set. Most organizations which use target costing select the straight line approach so that everyone is always working towards the same target cost.

EXAMPLE

A simplified example illustrates how a target cost (or target costs for different stages in a product's life cycle) can be set. The projections for a proposed new product A are as follows:

1 Total market life is five years for product A.

2 Forecast sales during the life cycle of product A are:

	Introduction (1st year)	Maturity (2nd and 3rd years)	Decline (4th and 5th years)
Unit price	£30	£25	£20
Sales volume in units	5 000	36 000	19 000

3 Total investment is £500 000

4 Required return on investment is 20 per cent per annum.

Alternative of one target cost for product A

		£
Sales	5,000 units × £30	150 000
	36,000 units × £25	900 000
	19,000 units × £20	380 000
		1 430 000
Cost of sales		?
Profit being 20 per cent of £500 000 × five years		500 000
By deduction cost of sales is		£930 000

Target cost = £930 000 / 60,000 units
= £15.50 per unit

Alternative of three target costs over life cycle of product A

	Introduction	Maturity	Decline	Total
Sales	£150 000	£900 000	£380 000	£1 430 000
Cost of sales	?	?	?	£930 000
Profit (*note 1*)	£52 000	£315 000	£133 000	£500 000
By deduction cost of sales is	£98 000	£585 000	£247 000	
Number of units	5000	36 000	19 000	
Target cost per unit	£19.60	£16.25	£13.00	

Note 1: Profit as a percentage of sales is £500 000/£1 430 000 which is approximately 35 per cent so that during the introduction stage profit is 35 per cent of £150 000 (i.e. approximately £52 000); during the maturity phase profit is 35 per cent of £900 000 (i.e. £315 000); and during the decline phase profit is 35 per cent of £380 000 (i.e. £133 000).

The above example results either in one target cost of £15.50 per unit over the entire product life cycle of new product A; or three different target costs during the three different stages of its product life cycle namely £19.60 during its introduction phase, £16.25 during its maturity phase and £13.00 during its decline phase. Whichever method is chosen, it is important that all involved are clear about the type of target cost being used.

PRODUCER'S, USERS' AND SOCIETY'S COSTS

By far the most common type of target cost is that of the producer, which takes into account only the costs incurred by the manufacturer. However, more organizations are realizing that this is perhaps too narrow a view of target costing and are also including the users' costs. For example, for a car buyer the costs of using the car include insurance, tax, fuel and maintenance. The general objective is to reduce the costs of using that product which may give the manufacturer a competitive advantage in the marketplace. Sometimes there is a trade-off between a manufacturer's target cost and users' target cost because customers may be willing to pay a higher initial price for a product if the future costs of using that product are reduced.

One or two organizations are taking an even broader view of target costing by including not only users' costs but also the costs of that product to society. These costs include environmental and social costs such as the costs of disposing of that product at the end of its life. The rest of this chapter will discuss target costing from the viewpoint of the manufacturer's target cost for illustrative purposes.

OBJECTIVES

Just as the objective of SVA is not only cost reduction, so the objectives of target costing also vary. Of course, cost reduction is usually a primary objective of target costing but other objectives include:

- improving quality
- satisfying customer needs better
- more timely introduction of new products.

The actual experience of using target costing has increased the importance of each of the above target costing objectives for many organizations. In particular, many organizations have found that target costing combined with the disciplined approach of SVA has speeded up new product developments. For some organizations this is particularly important because reducing the time to market gives a competitive advantage with higher initial prices and also greater long-term profits.

TARGET COST MANAGEMENT

The term usually is target costing but perhaps a better term is target cost management. Setting the target cost (despite its difficulties such as estimating a market price in the future) is relatively easy and the really difficult part of target costing is achieving the target cost set. This is why target cost management is a better term than target costing because the real issue is how you are going to manage to achieve the target cost. This is why the link between target cost management and SVA is critical.

One practical aspect of target cost management is who is going to be responsible for the target costing process and who is going to be involved in the team. Again there are similarities to the SVA team approach. The department given responsibility for the target cost management process varies from organization to organization and includes the following:

- design
- product planning
- accounting
- product technology
- research and development
- purchasing
- marketing
- manufacturing.

Similarly, the participants in the target cost management process include representatives from the above departments with one interesting addition from outside the organization, namely, suppliers. Mouritsen et al. (2001) report an interesting case study on target cost management involving suppliers. For many products the bought-in parts are a very significant percentage of total costs and therefore it is helpful, and indeed in some cases absolutely necessary, to have suppliers involved in the target cost management process. A critical element of target cost management is how to assign the target cost.

Assignment of target cost

Setting a target cost for a new or redesigned product or service is only the starting point in the target cost management process. The question is how to achieve this target cost. Usually this means breaking down the target cost into its sub-elements. Some organizations assign the target cost to blocks of components and then require the team of designers of such components to achieve that assigned target cost. Other organizations assign the target cost to the level of individual designers so that each designer has a specific target cost to achieve.

However, some of the most successful organizations assign the target cost to the functions of the product. This fits with the SVA process. A survey of Japanese organizations by Yoshikawa (1992) found that assigning the target cost to the functions of a product was the most common method of target cost assignment. One important advantage of assigning the target cost to the functions of a product is that this gives the designers more freedom in that they can decide how to achieve the necessary functions within the required target cost. In contrast, assigning the target cost to blocks of components requires the designers to use specific components; whereas the use of functions does not require the use of specific components.

The question remains, how do you assign the overall target cost to the different functions of a product? Perhaps the weakest method for assigning the target cost is on an arbitrary basis. Another method is the subjective basis where each member of the target cost team decides on the basis of their own experience how the target cost should be assigned and then an overall team basis of assignment is developed following discussion among the team.

However, it is generally recognized that the best method for assigning the target cost is on the basis of the views of the customers. Again this fits in well with the customer emphasis within the SVA process. This is also another way of bringing an external perspective into the costing system. An example of assigning a target cost of £300 to five different function areas based on the importance of each function from the customers' viewpoint is given in Figure 4.2. Market research would be used to determine the importance of each function from the customers' viewpoint.

Fig. 4.2 Assigning a target cost

Function	Importance of function from customers' viewpoint %	Amount of target cost assigned £
1	27	81
2	20	60
3	32	96
4	16	48
5	5	15
Total	100	300

Figure 4.2 shows that the customers considered function 3 to be the most important function and 'valued' it at 32 per cent and, with a target cost of £300, this meant that the target cost assigned to function 3 was 32 per cent of £300, namely, £96. The customers considered function 1 to be the second most important function at 27 per cent with a resulting assigned target cost of £81. In contrast to these functions 3 and 1, the customers considered function 5 to be the least important of these five functions and the target cost assigned to function 5 was 5 per cent of £300, namely, £15.

Achieving the target cost

It must be remembered that the whole process of target costing is only a structured approach in order to achieve certain objectives, and usually one of these objectives is cost reduction. After the target cost for a product or service has been set and then assigned, it is the responsibility of the designers to achieve the various assigned target costs so that the overall target cost is met using SVA. Generally, the earlier in the process that you assign the target cost, the better the overall results for two main reasons. First, the whole design or redesign process is generally better planned. Second, the designers have an assigned target cost at the earliest possible stage and, therefore, have more time to achieve the assigned target cost.

It is important that everyone is committed to achieving the target cost. It is very much a team effort. The person in charge of the SVA process is responsible for the overall co-ordination, so that any problem can be identified at an early stage and corrective action taken if necessary. For example, if the target cost for one particular function is not going to be achieved for some good reasons, the SVA co-ordinator will need to ensure that such an 'overrun' is offset by further savings elsewhere. Normally this is not a problem because for some functions designers will achieve an actual cost lower than the target cost.

Target costing is a cost management technique which focuses very much on the product or service from the market and customer perspectives. Target costing is an integral part of the SVA process. Strategic value analysis does not operate as successfully without target costing because target costing ignores the costs for existing products. As a result target costing influences costs well before the manufacturing process begins. The development, design and pre-manufacturing stages in a product's life cycle are when 80 per cent of costs are actually committed although not, of course, incurred. Target costing is a technique in the SVA process which allows cost management during the very earliest stages in a product life cycle.

PROBLEMS

Target costing is only a technique, and like all techniques it has its limitations and problems. One such problem is actually setting the target cost. It is not easy to look into the future and forecast the market price for a particular product, which is the starting point for the target cost calculation. Indeed, in some organizations this estimated market price leading to an initial target cost is only the start of a process which involves negotiation with all those involved. This is because the behavioural effects of setting a target cost are very important. If the team considers that a target cost is completely unrealistic and cannot be achieved, it is unlikely that the target costing and SVA process will be successful. It is critical that everyone in the team is committed to achieving the target cost, and this is where the negotiation process can be helpful.

A second problem is linked to this first problem of setting the target cost, and concerns the assumptions underlying the target cost. These assumptions include, for example, the competitors in the future. Such competitors may include both existing competitors and new entrants into the market. A second assumption relates to future technological developments. For example, if you were setting a target cost for a wind-up wristwatch in the past, would you have anticipated the new technological development of quartz watches? A third assumption concerns customers' preferences in the future. Again, these have to be estimated. The overall point is that the assumptions underlying any target cost mean that it is by no means an exact measurement. A target cost is the best estimate at a particular point in time give certain assumptions. A target cost is in reality a ballpark estimate.

A third problem is whether to set one fixed target cost or a series of stepped target costs over the early stages (planning, design and manufacturing) of a product's life cycle. Perhaps the most important point is that everyone understands what type of target cost has been set.

A fourth problem concerns the exact costs to be included in the target cost. It can be argued that at least all the manufacturer's costs should be included in the target cost. However, is a target cost also being set for the costs incurred by users and, if so, is there to be any trade-off between the manufacturer's target cost and the users' target cost?

If all the manufacturer's costs are being covered by the target cost set then a fifth problem is determining the volume of production on which the unit target cost is based. This is particularly important for the development costs and overhead costs. If a traditional approach to overhead absorption is followed (such as using labour hours or machine hours), this is a particularly serious problem. However, the problem can be lessened somewhat by using an activity-based approach to overheads (see Chapter 5 for further details). Again everyone involved in the target costing process needs to be clear about which overhead approach is being used and the volume of production on which the unit target cost is based.

The sixth, and perhaps the most important, problem is the target cost management, i.e. how the target cost is actually going to be achieved. After the problems of setting and assigning the target cost, there is the target cost management stage leading to the SVA process freeing up the creativity of all those involved.

ADVANTAGES

One big advantage of target costing is moving away from the internal focus of traditional costing (with its build-up of materials, labour and overhead elements into a unit cost) to an external focus (bringing market information into the costing system). The starting point is a future market price working back to a target cost.

A second advantage is that target costing is a useful technique which supports the SVA process during the development and design of new products (or services) or during the redesign of existing products (or services). With such a high percentage of costs being committed during the development and design stages of products (or services), it is important that not just development and design personnel are involved in the decisions taken during these stages. Target costing and the SVA process allow managers from other areas (such as accounting, manufacturing, marketing and purchasing) to be involved in such development and design decisions.

A third advantage of target costing is that it can aid faster new product development. Target costing together with SVA gives a structured team approach to the design of new products (or services). Very often this structure improves communication and many organizations have found that this common team objective of achieving a target cost has speeded up both the development and design stages for new products. Furthermore, the experience of many organizations is that the target costing and SVA team approach reduces the number of problems at the pre-production and production stages because with production being involved from the very beginning of the process, most production problems are anticipated and solved during the design stage.

A fourth advantage of target costing is that it links customer requirements very closely to the design process. For example, the assignment of the overall target cost to the individual functions of a product is best done on the basis of customers' views on the importance of each function. Such customers' views can be very different from designers' views but the customers' views have priority in the target costing and SVA process.

A fifth advantage of target costing is that some organizations have found that it improves the quality of both new and redesigned products. Again the improvement in quality can be traced to two aspects, namely, the structured SVA approach and the team approach. Throughout the development and design stages, managers other than designers (for example, manufacturing and purchasing) can have an input into the process and make suggestions for improvement. Experience has shown that such suggestions can improve the quality of the final product without increasing its cost.

CONCLUSIONS

Target costing is an integral part of the SVA process. Target costing is revolutionary when compared with traditional costing with its internal emphasis and its build-up of costs based on material, labour and overhead elements. In contrast, target costing has an external emphasis starting with a future market price and working back to a unit cost by deducting the required profit margin.

The really difficult part of target costing is actually achieving the target cost, and this is where target cost management leads into the SVA process. Like SVA, target costing is a technique with a structured approach for an interdisciplinary team.

Different organizations have individuals from different departments leading the target costing team because the leader basically fulfils a co-ordinating role for the members in the target costing team.

Another important element of target costing which links with SVA is the assignment of the overall target cost to the individual functions of a product based on customers' views of the importance of each function. Target costing has both a market and a customer emphasis. Target costing is an important technique in the SVA process of the design of new products (or services) or the redesign of existing products (or services).

5

Cost tables

DEFINITION

Cost tables are of great assistance during the SVA process. Cost tables have been in existence for many years. For example, Sato (1965, p. 51) defined cost tables as 'a measurement to decide cost and to be able to evaluate the cost of not only existing products but also future products at the very beginning of the design process'. In contrast traditional costing systems have concentrated on costing existing products and services, whereas cost tables have been designed to answer 'what if' questions from managers, designers and others. For example, when designing a new chair what are the cost implications:

1 if a different material is used;
2 if the shape of the legs is changed;
3 if the height of the back is increased.

Cost tables can provide the answers to such 'what if' questions.

Cost tables are databases of detailed cost information based on various manufacturing variables. In effect these are 'cost drivers' for direct costs – in other words direct costs are driven by factors in addition to volume of production. Of course, volume of production remains an important cost driver but other factors also influence costs and cost tables take these other factors into account. For example, one driver of the cost of a motorcycle is the cubic capacity of the engine. Similarly, one driver of the cost of a conveyor belt is its length.

Another aspect of cost tables which distinguishes them from traditional costing systems is the fact that cost tables include information from both within and outside an organization. Traditional costing systems usually include mainly internal information but cost tables include both internal and external information. For example, cost tables will include external information about new materials, new machines and new manufacturing processes. Obviously, this external focus encourages the compilers of cost tables (usually accountants) to have an external focus and to be aware of the latest developments which might affect their organization. One source of such external developments is the local university. This external focus of cost tables fits well with the SVA approach with its external focus on the market and customers.

Cost tables are a useful technique both for the target costing and SVA processes during the design of new products or the redesign of existing products. The cost implications of different design alternatives can be worked out relatively easily instead of doing a one-off exercise for each alternative design. Cost is, of course, only one factor in the design process but cost tables mean that the cost factor can be considered from the very beginning of the design process. With cost tables, very often one or two design alternatives can be eliminated during the earliest phases of the SVA process and this can save valuable design time which instead can be spent on other more feasible design alternatives.

APPROXIMATE COST TABLES

(see Yoshikawa et al., 1990)

The are two main types of cost tables, namely, approximate cost tables and detailed cost tables. As the names indicate the approximate cost table is a simplified version of the detailed cost table. In other words, you develop an approximate cost table first and, if you find it useful and require more detail, you can then develop the approximate cost table into a detailed cost table. In essence, the approximate cost tables has fewer cost drivers and less external information than the detailed cost table.

An approximate cost table is particularly helpful during the early stages of the design process when you are assessing different design alternatives and approximate costs (i.e. 'ballpark costs'). An example of an approximate cost table is given in Figure 5.1 which shows a cost table for drilling based on different depths of hole and different types of materials used.

Fig. 5.1 Example of approximate cost table

Note: This is a illustrative example and the costs are hypothetical.

Activity: drilling
Equipment: mark 3 power drill
Volume: × units per annum

Type of material \ Depth of hole	3 inches				5 inches				7 inches			
	Mat. £	Lab. £	Oh. £	Tot. £	Mat. £	Lab. £	Oh. £	Tot. £	Mat. £	Lab. £	Oh. £	Tot. £
Plastic	5	2	3	10	7	5	5	17	8	7	8	23
Steel	9	2	2	13	10	2	2	14	12	4	5	21
Aluminium	10	2	2	14	11	3	3	17	12	3	4	19

Figure 5.1 gives a simplified view but it shows plastic giving the lowest cost of £10 for the 3-inch hole, steel having the lowest cost of £14 for the 5-inch hole and aluminium having the lowest cost of £19 for the 7-inch hole. In Figure 5.1 the number of units per annum is significant for the overhead figures, and this problem of overheads in cost tables will be discussed later under the heading of the activity-based approach.

DETAILED COST TABLES (see Yoshikawa et al., 1990)

Approximate cost tables provide useful information during the very early stages of the design and SVA process but, as the name suggests, detailed cost tables give much more information and have various uses including design, purchasing, subcontracting and production costing. For example, instead of relying on competitive tendering for purchasing, detailed cost tables give a good estimate of what the cost should be if the best materials, machines and production processes are used. Such information provided by detailed cost tables can be very useful during purchasing or subcontracting negotiations. Figure 5.2 gives an example of a detailed cost table.

Fig. 5.2 Example of detailed cost table

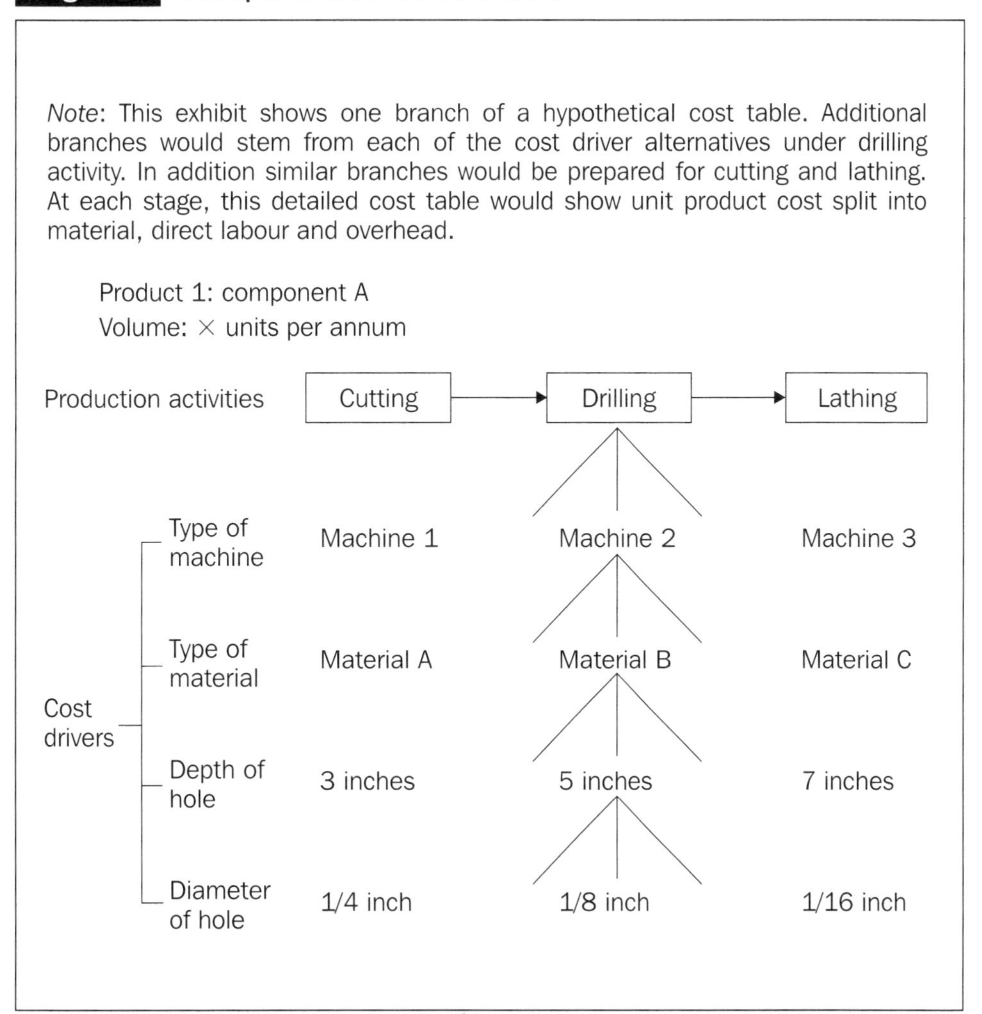

Figure 5.2 expands on the approximate cost table of drilling in Figure 5.1, which included information only on the type of material and the depth of hole. The

detailed cost table in Figure 5.2 includes more variables, among these being the type of machine and the diameter of the hole. Originally, cost tables were in a paper format but now, of course, they are computerized and are, in effect, a database which can provide answers to the 'what if' questions. For example, the detailed cost table in Figure 5.2 could give the cost of drilling using machine 3 with material C for a 5-inch deep hole with a diameter of a quarter of an inch.

Figure 5.2 illustrates how the permutations increase as the number of variables increase. The maintenance and updating of such a detailed cost table is a major task with not insignificant costs involved. For example, in a factory with 1000 employees there could be two accountants working full time on maintaining and updating that factory's cost tables. These cost tables can also be combined with computer-aided design so that designers can see the cost implications of different design alternatives at the press of a button. This is obviously helpful during the SVA process.

PARTS OR FUNCTIONS

The two types of cost tables (approximate and detailed) can be based on either the physical parts or the functions of a product. The functions of products (such as make mark for a pen) are discussed in Chapter 2. Cost tables based on product functions will have a wider application than cost tables based on the physical parts of a product. Functions are at the heart of SVA and enable managers to thinks about products (or services) in more abstract terms.

Whether parts or functions are used as a basis for cost tables, there are significant start-up costs in establishing such cost tables. There is a learning curve about knowing what to include in cost tables. If you are developing cost tables from scratch, it is probably better to base these tables on the functions and to begin with approximate cost tables. After some experience with approximate cost tables, based on functions, it can be decided whether it is worth the additional cost of developing detailed cost tables.

COST MANAGEMENT

Cost tables play a useful role in cost management during the SVA process. The external information incorporated into the cost tables can lead to questions such as:

1 Why is a new production process not being used?
2 Why is a new machine not being used?
3 Why are labour costs so high?
4 Why is a different material not being used?

During the SVA process cost tables can also help to identify where additional expenditure might be worthwhile from the viewpoint of customers to generate more profit. It is important that accountants and managers consider not only cost reduction but also where additional expenditure can increase the organization's profit. Cost management is about generating more profit and not just about cost reduction.

A major advantage of cost tables is that such tables give managers a better understanding of what actually drives costs. Volume remains a major cost driver but cost tables reveal that there are also a number of other drivers of material, labour and overhead costs. By understanding these other cost drivers, managers can manage costs better. One such area is that of overhead costs, and the activity-based approach to overheads can be incorporated into cost tables to help the SVA process.

ACTIVITY-BASED APPROACH TO OVERHEADS

If overhead costs are a relatively insignificant percentage of total costs (say less than 5 per cent), then it does not matter very much how overheads are incorporated into cost tables. However, if overheads are a significant percentage of total costs, then it is worth considering whether the extra costs involved in the activity-based approach might be worthwhile.

In the traditional method, overhead costs are collected in departments such as purchasing, traced on some basis to production departments and then linked to individual products by using an overhead absorption basis such as direct labour hours or machine hours. The basic problem with such overhead absorption bases is that they vary with the volume of output and yet many overhead costs are relatively fixed and do not vary with the volume of output.

In the activity-based approach, overhead costs are collected in terms of activity-based cost pools such as material handling. A second difference from traditional overhead costing is that activity-based cost driver rates (such as cost per material movement) are used to relate the activity costs to individual product lines. Such cost driver rates do not vary with the volume of output. The two main differences between the traditional approach to overhead costing and the activity-based approach are illustrated in Figure 5.3.

Determining activities

At the core of the activity-based approach is determining activities. This process involves interviewers (managers from within the organization aided by consultants in some cases) asking managers and other employees what they do. Often, at first very detailed tasks or sub-activities will be identified but, with further interviews and analysis, the main activities will emerge. Examples of activities include the following:

Fig. 5.3 Comparison of traditional and activity-based approaches to overhead costing

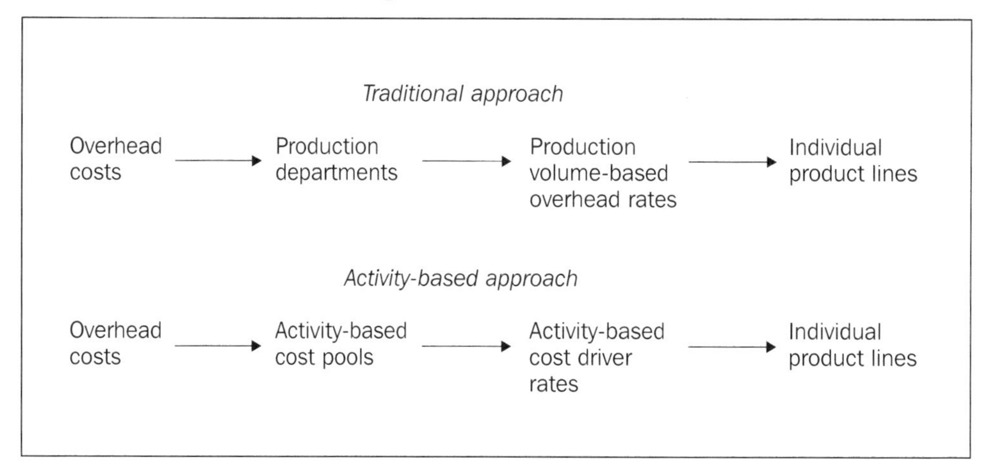

- purchasing (note that this activity goes well beyond the purchasing department to include purchase requisitioning, checking financial status of suppliers, receiving and paying suppliers)
- material handling
- training
- accounting
- shipping
- marketing
- serving customers.

An important feature of many activities is that they cross departmental boundaries and in some organizations are described as processes. After selecting the main activities, the next step is to determine the costs involved in each activity namely the activity cost pools. It is vital to reconcile the total overhead cost for the financial ledger with the total in all the activity cost pools to check that the activity-based system has included all the overhead costs.

Selecting activity cost drivers

After determining the activities and the activity cost pools, the next step is to select appropriate cost drivers for each cost pool. The aim is to identify the main driver or cause of costs in each cost pool. Examples of cost drivers include the following:

Activity	Cost driver
Purchasing	Number of purchase orders
Material handling	Number of material movements
Customer liaising	Number of customers

The cost drivers are usually volume based but are not usually based on the volume of production. This is an important difference from traditional overhead costing where the number of machine hours or direct labour hours are very closely related to the volume of production.

Overhead costs per unit

A good cost driver must also be measurable and often a cost driver may be the best measure of the capacity of that activity. After selecting the activity cost drivers, cost driver rates can be calculated for each cost pool namely:

$$\frac{\text{Activity cost pool}}{\text{Cost driver}}$$

For example, the activity of material handling costs £10 million and the cost driver for the activity of material handling is the number of material movements with 100 000 material movements per year. The cost driver rate for material handling would be £10 million/100 000 = £100 per material movement. If 1 million units of product A are manufactured each year with 20 000 material movements per year for product A, then the material movement cost per unit for product A would be:

$$\frac{20\,000 \times £100}{1 \text{ million}} = £2$$

Of course, the total overhead costs remain the same but the activity-based approach gives a different analysis of the overheads to the various product lines leading to different unit product costs from the traditional approach to overhead costing. A typical result of applying activity-based costing is that it reveals that under traditional unit product costing, large batch products tend to cross-subsidize small batch products. Basically this happens because traditional overhead costing uses mainly volume of output overhead absorption bases such as direct labour hours and machine hours, whereas such bases are the exception under the activity-based approach.

The activity-based input to cost tables is very much the overhead element of unit product costs. However, the activity-based approach also has a role to play in the SVA of overheads but this topic will be discussed in Chapter 8. The activity-based approach helps to reduce the seriousness of the overhead problem in the unit product costs in the cost tables but, of course, the problem of incorporating overhead costs in a unit product cost remains. There are also other problems associated with the use of cost tables.

PROBLEMS

Starting from scratch

Building cost tables from scratch is very time-consuming and is very much an investment for the future. However, the cost of building such cost tables has discouraged some organizations. Other organizations have decided to begin with approximate cost tables because it is much more time-consuming to develop detailed cost tables because of the greater number of cost drivers involved.

External Data

Another problem is finding and incorporating appropriate external data. It is often difficult to decide what external data to include in the cost tables. For example, developments such as new materials or changes in technology may be excluded from the cost tables because such developments may be considered irrelevant. The basic guideline is if in doubt include such data.

Maintenance of cost tables

In addition to the time investment in constructing cost tables, there is an even bigger long-term investment in maintaining and updating an organization's cost tables. Maintaining and updating cost tables so that they continue to be useful is a never-ending task.

Amount of detail

Another problem is how much detail to include in the cost tables. For the approximate cost tables it is a question of selecting the most important cost drivers. It is perhaps even more difficult to decide what to include in the detailed cost tables. Basically it is a cost-benefit decision. The question is do the benefits of including extra data exceed the costs of incorporating such data in the cost tables. The circumstances of each organization will affect such a decision.

CONCLUSIONS

Despite the above problems associated with cost tables, it is well worth considering whether to develop cost tables which will be extremely helpful during the SVA process. Cost tables are particularly useful during the design of new products or the redesign of existing products. Cost tables mean that the costs of different design alternatives can be compared easily.

During the redesign of existing products, the main advantage of cost tables is very often not the existing cost information for that product but, rather, the other information in the cost tables. Such other information in the cost tables may stimulate the design process. For example, the external information in the cost tables may include details about a new material or a new production process which could be used for the existing product.

Cost tables are also useful for estimating what the cost should be for a component to be purchased from a supplier. Similarly, if work is to be subcontracted, the cost tables can be used to estimate what the cost should be before the subcontractor quotes for the work.

Indeed, if you were building a cost system from scratch, it would seem reasonable to construct such a system to answer the 'what if' questions from managers. In particular cost tables based on functions are geared to the future rather than the past and, in addition, such cost tables also have an external emphasis.

It is very difficult to build detailed cost tables from scratch but much easier to begin with approximate cost tables. Some organizations consider that such cost tables give them a competitive advantage. It is still possible to conduct SVA without cost tables, but such cost tables make the SVA process much easier.

BISHOP BURTON COLLEGE
LIBRARY

6

Kousuu

The success of SVA activities is dependent on relatively quick assessments being made of the impact of new functions, function modification and function elimination both on product or service prices (and hence on revenues) and on costs. In the case of the latter, exact money measurements would be complex and could delay a team's progress. This is especially true where the team does not have accounting members. Consequently it has become common in Japan to initially assess the changes suggested by the SVA teams in terms of their expected impact on resource consumption. This is done in non-financial terms through the use of a series of resource consumption rates which are a familiar part of the work environment of those participating in product-centred SVA work. Thus the impact of work changes on these rates are more readily determined by the SVA team. These rates are known as Kousuu and this chapter explains how they are constructed and used to support SVA.

THE NATURE OF KOUSUU

Kousuu are rates of resource consumption expressed in physical terms. They cover all the major resource elements of conversion and support cost, and typically are based on measures of direct or indirect labour time and machine time related to the production factors which constitute the organization. A complete set of Kousuu thus provides a detailed profile of all of the conversion and support activity undertaken in the firm. Kousuu can be based on production processes, work cells, machine and service functions such as maintenance and materials handling. This type of information can be usefully presented in various ways. For example, with an input object focus it can be designed to represent the time distribution of the various resources constituting a production line, a shift, or a factory for any specified period of time. However, by focusing on an output object, Kousuu may also be expressed in terms of the various time components of the work required to produce one unit of final product. In this latter form it is known as Gentani. A Gentani therefore profiles the pattern of conversion work resource consumption by individual product lines. In this form the information is particularly relevant to product-centred SVA teams. In addition, it provides a valuable working performance measure in its own right. Finally, to accommodate the financial dimension, a charge rate can be computed for each Kousuu based on the cost of the resources which contribute to the labour and/or equipment and service input of the relevant activity. This can then be used to convert the Kousuu work times into costs which can be applied to all the above types of cost object. In Japan, Kousuu are extensively used in the manufacturing sector and their design and operation are widely referenced in applied texts (e.g. DES, 1989: GBD, 1991).

One of the strengths of the system of Kousuu is the level of detail which it captures and feeds back to management. It thus provides a checklist for the SVA team to identify how their decisions will impact on resource consumption and hence cost. If presented to identify areas of potential waste, it may also initiate ideas for change for the SVA team. For example, Figure 6.1 contains an illustration of a Kousuu based on labour working hours for a particular production process for a specified time period. From left to right there is a hierarchical decomposition which first classifies working time into that which adds value (basic working hours) and that which does not (line management hours) and then a further subdivision is made into direct or support work and, finally, a segmentation into the detailed activities which constitute each of these components.

This pattern of decomposition is one of the advantages of Kousuu as it highlights how resource is consumed and in so doing it facilitates the identification of non-production time and guides the SVA team on how operational improvement might be achieved.

These attributes are further enhanced by establishing responsibility linkages for Kousuu components. These may be linked to SVA team composition. For example, all the basic working hours will typically be the responsibility of the production engineering department. They will attempt to devise new designs, new methods and work support services which will improve the utilization of direct work time. The line management hours will be the responsibility of the production department who will attempt, over time, to reduce and/or eliminate the non-value added work from this component.

Frequently, for performance measurement in SVA activities, Kousuu are expressed in the form of Gentani. These would be computed in the above example by dividing the columns in Figure 6.1 by the number of finished products manufactured. This type of analysis highlights the significance of the various Kousuu components in a way which is directly related to actual achievements in respect of meeting the cost targets set for the team. Indeed, targets may be set for some teams (where the members are more likely to relate to non-financial measures) in terms of Kousuu savings. They also allow progress towards meeting a cost target to be monitored initially in non-financial terms such as the volume or proportion of net working hours or incidental working hours saved by a suggested functional modification.

One notable feature of Kousuu is the segregation of work time into its value added and non-value added components. Identification of non-value added hours provides a particular focus for cost reduction effort. It also enables the use of the ratio of value added to non-value added time components which represents another important attention-directing Kousuu-based performance measure for the SVA team. Indeed, it is a key test of most Japanese cost management policies and initiatives that they will impact favourably on the organization's Kousuu. Reduction of Kousuu is a key indication of cost effectiveness as it provides evidence that costs will fall. Without this type of evidence the value of any new developments can be called into question as they will lack managerial credibility. Thus the analysis of any new initiative's impact on Kousuu will be a crucial element of the case for its adoption. The translation of this impact into cost savings can be done subsequently, once the real impact of the change has been identified and measured in Kousuu terms.

Fig. 6.1 Working hours for Kousuu

Type of working hours			Activities
Working hours (WH)	Basic working hours (BWH)	Net working hours (NWH)	1. Machine loading and unloading 2. Working manually or operating machines 3. Supplying parts daily 4. Washing processed parts and finished products 5. Measuring processed parts and finished products
		Incidental working hours (A)	1. Walking between process 2. Dressing parts and products 3. Loading parts on automatic machines 4. Adjusting machine tolerance 5. Checking size of processed parts and finished products by random sampling 6. Cleaning machines
	Line management hours (LMH)	Incidental working hours (B)	1. Turn on and off the main switches 2. Preparing parts for manufacturing 3. Preparing and checking tools 4. Checking machine and supplying oil 5. Cleaning floors 6. Warming up machines 7. Holding preliminary meeting and making contact with workers 8. Checking blueprint
		Incidental working hours (C)	1. Changing cutting or grinding oil 2. Changing running or lubricating oil
		Set up hours (SU)	1. Changing fitting and fixing tools 2. Changing manufacturing tools
		Artificial delay hours (AD)	1. Relating to abnormal shop floor works 2. Relating to factory management 3. Relating to personal issues
		Waiting hours (W)	1. Waiting for manufacturing parts and products

Source: Yoshikawa et al. (1997)

CONCLUSIONS

Kousuu do have to be used with care as they focus on the sensitive issue of work time. Their detailed analyses put employees 'under the microscope' and, particularly in the West, their very existence may lead to human relations problems. Moreover the elimination of all 'slack' time can increase work pressure and lead to a loss of employee freshness, creativity and morale.

With this caveat their potential value is considerable. In respect of SVA it plays two important and complementary roles. First, it provides a measurement basis which allows proposed changes to be translated into resource savings in a manner meaningful to those engaged in the SVA team. Second, a system of Kousuu monitoring and reporting (Yoshikawa et al. 1997) can help direct the attention of SVA teams to issues and areas where they can suggest ways of improving their firm's cost effectiveness.

7

SVA case study

INTRODUCTION

This chapter contains a description of how SVA was used by a large Japanese transport sector corporation to improve its cost effectiveness. To preserve anonymity the name of the corporation and the figures used have been altered.

SVA ACTIVITY HISTORY

Edony Corporation (from now on called Edony) is one of the largest subcontractors in the automobile industry. It supplies a variety of automobile parts for major Japanese automobile companies. One of their key operational strengths has been their dedication to the use of SVA activity in order to meet their objective of manufacturing high-quality products with a low price.

Edony has devoted more than 8000 hours (this would translate to over 48 000 person hours) every year to SVA team activity (see Figure 7.1). This does not include personal SVA activity hours which are three times more than those of the SVA team activity. The corporation plan to spend from 30 000 hours to 40 000 hours in total this year. Personal hours will greatly inflate this time commitment.

Fig. 7.1 SVA team activity hours

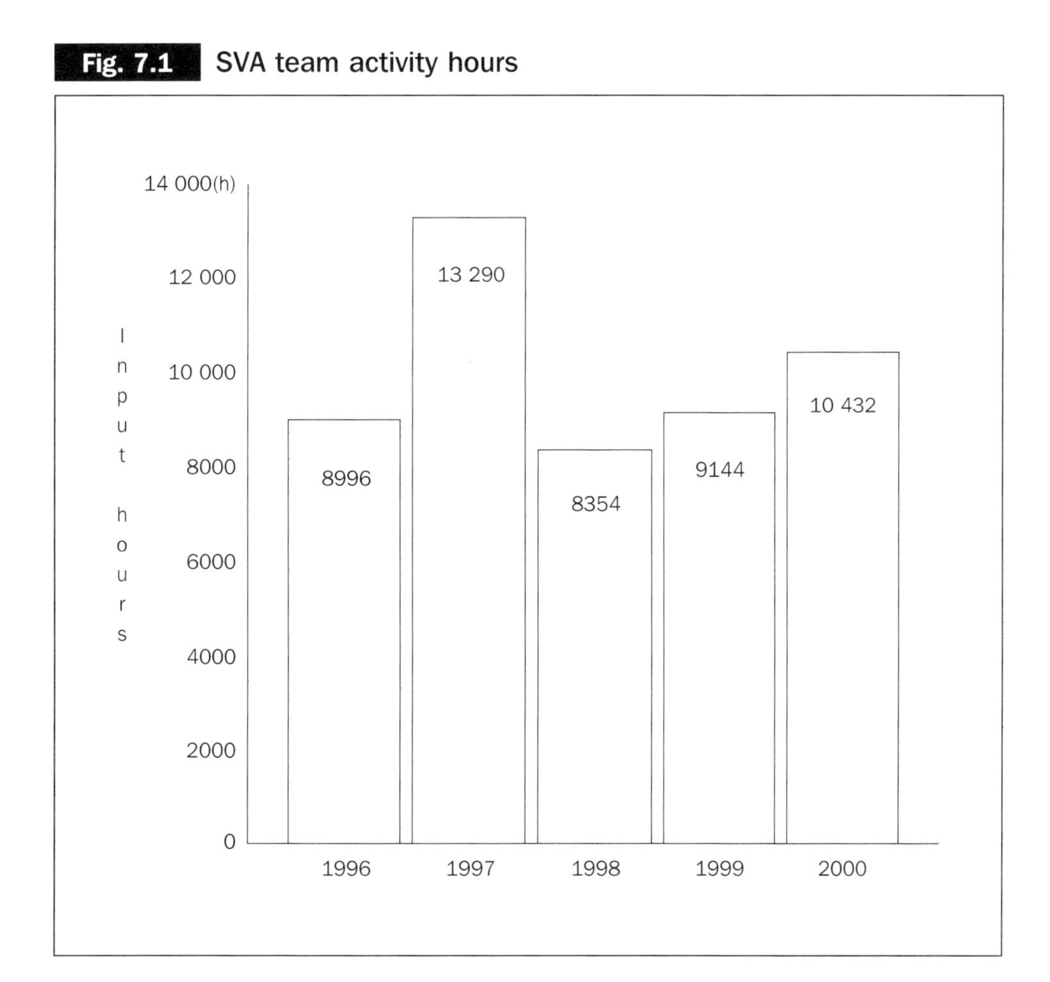

History of SVA education

Edony has two types of SVA education programmes. One is for beginners and the other is an advanced one. The first is for new employees who have to participate in SVA team activity and learn SVA as on-the-job training (OJT). Figure 7.2 shows the number of new employees who learned SVA by OJT. They have educated 1278 new employees in SVA so far (see Figure 7.2).

Fig. 7.2 Number of new employees who are educated in SVA by on-the-job training

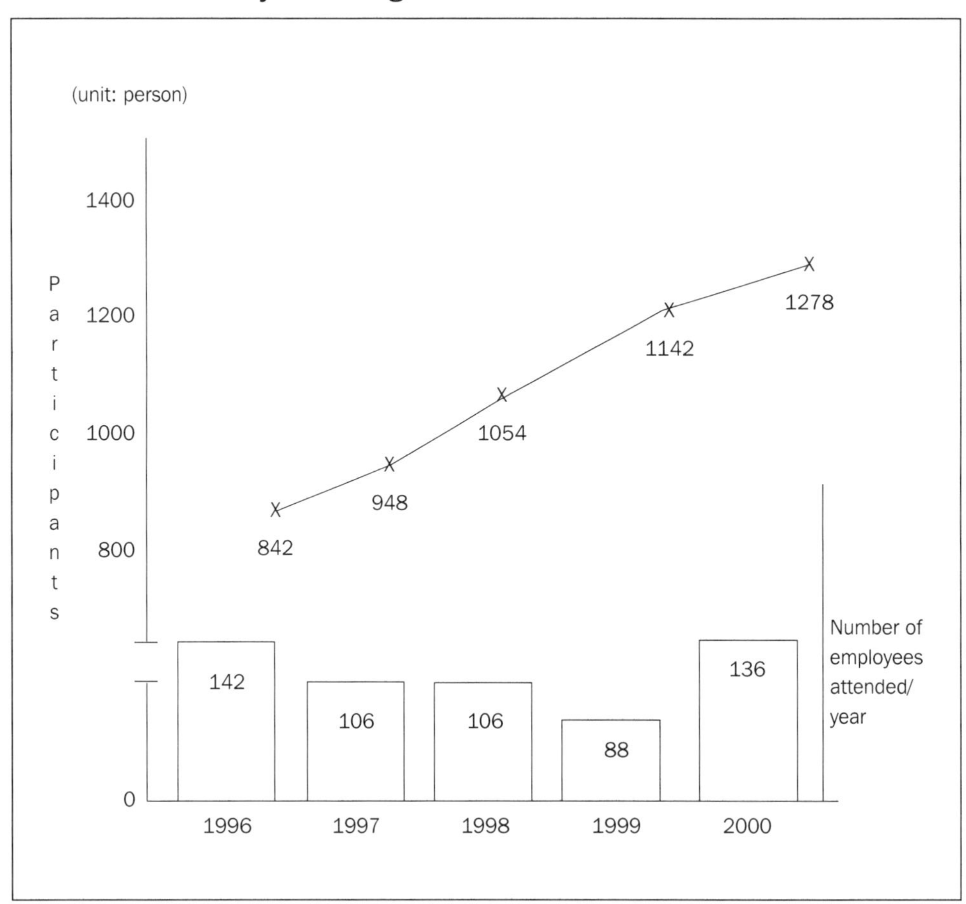

The advanced course is for senior managers who are expected to be the team leaders of SVA activity. They have to learn a functionally oriented product (service) improvement, design and manufacturing way of thinking. They have to learn how to define the functions of the product (or service), and improve their product (service) based on its functionality. Figure 7.3 shows the accumulated number of senior managers who have attended this programme.

Fig. 7.3 Cumulative number of senior managers who have attended the advanced SVA programme

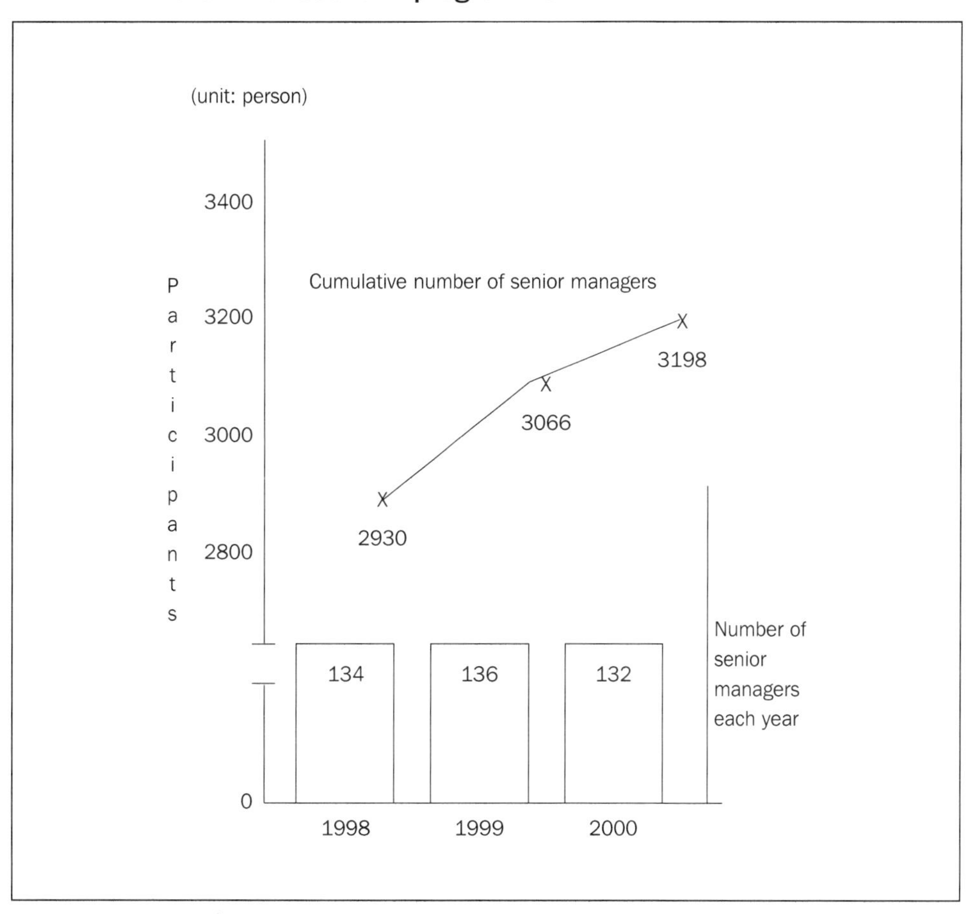

History of the results of SVA activity

After learning the basic idea of SVA, i.e. 'What is the function of the product (service)', 'What is the purpose of these functions' and 'Is there any alternative mean for the purpose', participants apply the educational results to their small team activities and increase the quality and quantity of their activity. As a result they have been successful in raising the number of improvement suggestions which currently totals more than 300 000 (cumulative number) for the last two years. Specifically they received 192 000 improvement suggestions in the last year. Figures 7.4 and 7.5 provide details of how suggestions have accumulated over the last five years. This reflects the high intensity of SVA activity in Edony.

Fig. 7.4 Number of cumulative improvement suggestions in the company

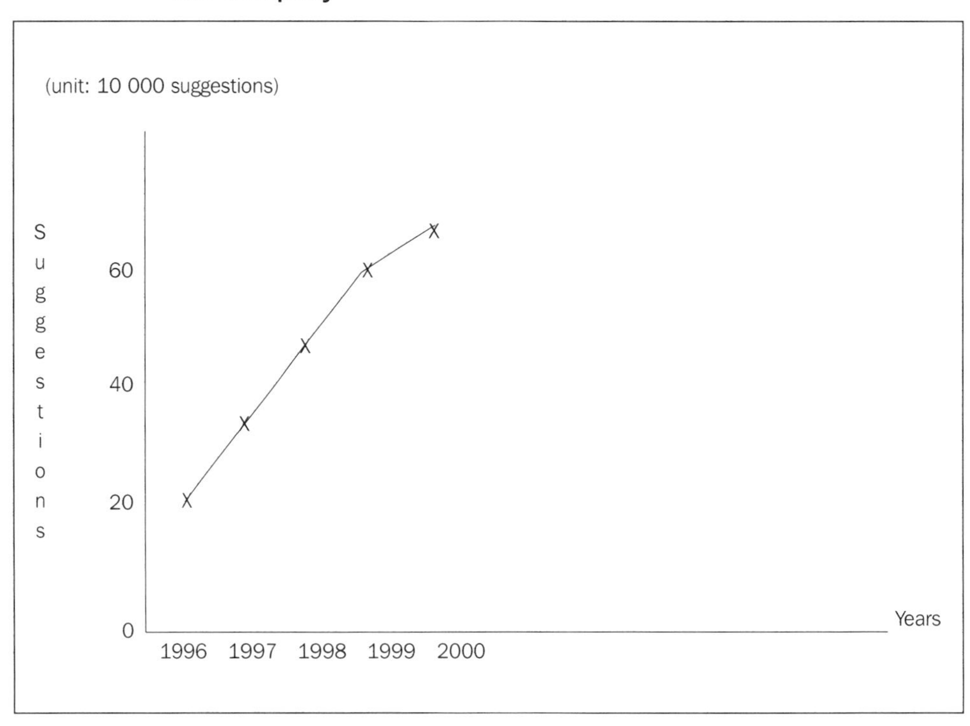

Fig. 7.5 Number of cumulative improvement suggestions by person

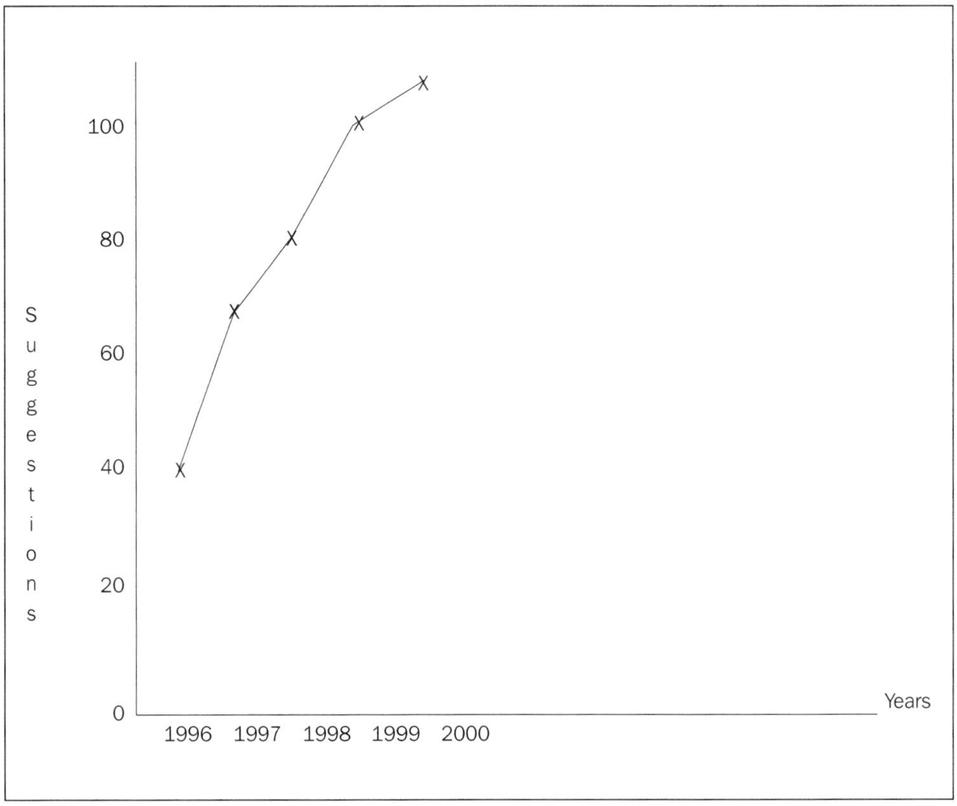

Targets for Kaizen (continuous improvement) activity are established in terms of Kousuu (work time targets) and costs based on a divisional profit plan. Details of Kousuu are given in Chapter 6. Once the target is set, it is then broken down into departmental targets. The staff in each department then start to develop SVA activity to achieve the target. The results of their activity is subject to ongoing evaluation as an achievement rate (i.e. actual activity results versus target) and/or as a sales rate (amount of cost savings/sales) at the end of each period. Figure 7.6 shows the sales rate and amount of cost savings for Edony. This illustrates the increasing financial success of SVA activity over the period.

Fig. 7.6 Sales rate (amount of cost saving/sales) and amount of cost saving

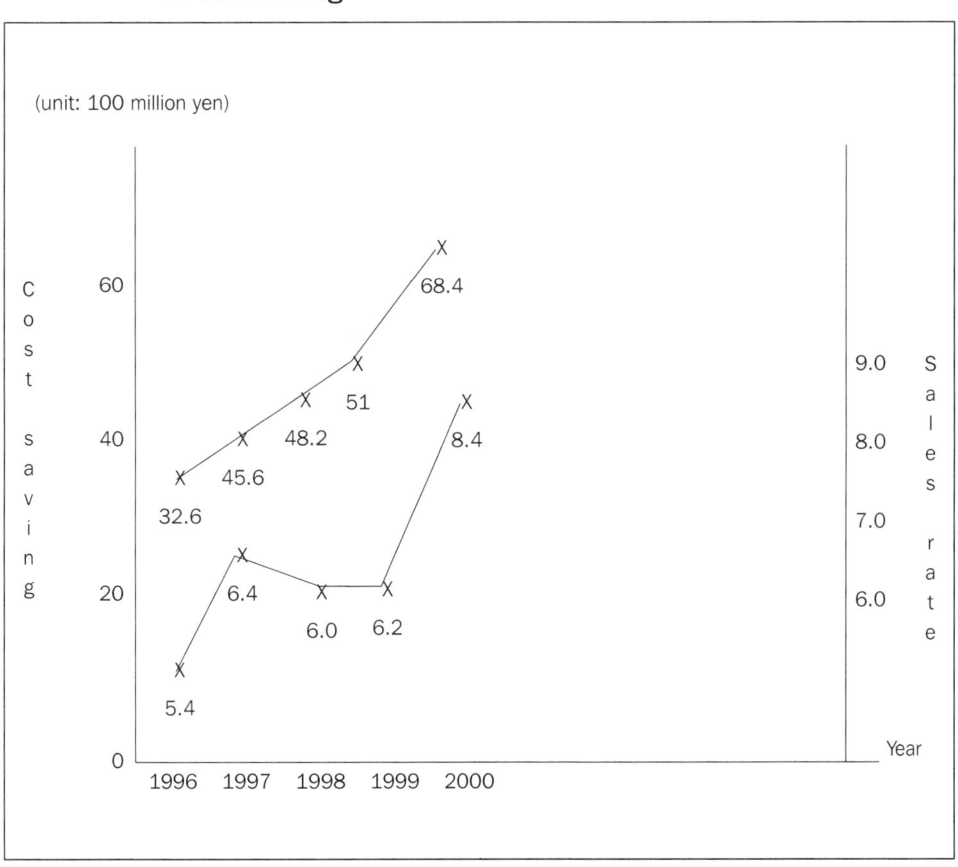

COST REDUCTION METHODS

In addition to changes in the functional structure of the product, the efforts of the SVA team are intensified by screening each function with a number of cost reduction possibilities. These are considered, in turn.

1 Can the type of material used to achieve a function be changed to a lower cost substitute with no significant impact on customer satisfaction? This analysis will depend heavily on the interaction between SVA team members from procurement and product engineering.
2 Can the material used enable a reduction in product weight/bulkiness? Even if this requires a more expensive material to be purchased, it may be worthwhile if it enables cost savings to be made at other stages of the production process.
3 Can the number of parts/components used in the product be reduced even further? Reducing parts can enable the assembly process to be simplified and can also have a beneficial cost impact on procurement and supply chain management.
4 Can the design/operation of the function be simplified? This makes the production process easier to learn and to carry out without mistakes. Failures can therefore be reduced, yield rates increased and quality improved.
5 Can the production processes required for a function be done by a less expensive type of labour and/or can they be automated? If so, conversion cost can be reduced.
6 Can any cost savings achieved be used constructively? For example, can existing functions be improved or extra functions added from identified savings. This provides a 'no lose' approach to the enhancement of product value which can justify higher selling prices.

From all the above questions the cost aspects and the functionality of products can be jointly analysed and action taken to enhance the product's cost effectiveness. All this, however, is dependent upon obtaining and amending the visual representation of the product which is manifest in its functional family tree.

8

SVA of overheads and services

MANAGING OVERHEADS

For most organizations the management of overheads is a problem area. There are relatively few techniques which managers find helpful in this area. This section provides an overview of two existing techniques namely zero-base budgeting and activity-based management which are used to manage overheads. The reason for discussing these two techniques is that SVA uses elements of both zero-base budgeting and activity-based management to develop a third relatively new approach to the management of the delivery of overhead services to customers, namely, to managers in other departments in the organization.

Zero-base budgeting

Zero-base budgeting was introduced by Texas Instruments in the late 1960s. Phyrr (1970) wrote about zero-base budgeting and claimed that there were more than 100 American organizations using this technique at that time. The State of Georgia was using zero-base budgeting and when its Governor, Jimmy Carter, became President of the USA, he ordered all federal government agencies to adopt zero-base budgeting. However, both public and private sector organizations have used zero-base budgeting.

Zero-base budgeting begins with a base of zero rather than with the current year's budget or actual results. This means that priorities must be decided not just for new initiatives but also for all ongoing activities. Equal attention is given to both existing and projected activities. Resources are not allocated to existing overhead activities just because these have always been funded in the past. The basic assumption is that the organization is starting from a zero base. Each overhead activity needs to be reconsidered before any resources are allocated to it.

Another feature of zero-base budgeting is that it begins with what are technically called 'decision units' which are the lowest level of budgeting units in an organization concentrating on activities. These decision units are usually different from the existing budget department and very often these sub-unit activities cross existing departmental boundaries.

Each decision unit prepares a set of 'decision packages' covering both existing and future activities. These decision packages are then evaluated by managers in terms of cost-benefit to the organization. The result is that some existing activities may cease because they are ranked lower than some new proposed activities.

The basic problem with zero-base budgeting is that the managerial time taken to produce the basic information for a zero-base budgeting system is much greater than in a traditional budgeting system because managers need to assess all activities (not just new activities) because they are starting from a zero base. Since its development in the late 1960s, many organizations have tried zero-base budgeting but today relatively few have full zero-base budgeting systems. However, some organizations have used zero-base budgeting on a one-off basis and others have applied it to selected parts of their organization where problems are being experienced. The two main contributions of zero-base budgeting are:

1 it has raised questions about the traditional incremental approach to budgeting;
2 it has made managers realize that existing activities must be examined as closely as any new activities.

Activity-based management

Activity-based management (see also Chapter 5) is concerned only with overheads (both production and non-production) and not with direct materials, direct labour or direct expenses. There are three main aspects of the activity-based approach:

1 activities;
2 activity cost pools;
3 activity cost drivers.

These three aspects of the activity-based approach are discussed in more detail in Chapter 5.

After identifying the overhead activities, cost pools and cost drivers, the activity-based management can begin. One aspect of activity-based management is simply the different analysis of the overhead costs. For example, a traditional analysis of the costs of the overhead department of purchasing is show in Figure 8.1.

Fig. 8.1 Traditional analysis of costs of purchasing department

	£000
Salaries	350
Travel	150
Depreciation on equipment	100
Other expenses	50
Total	650

In contrast, the activity cost of purchasing (including activities outside the purchasing department) is £900 000. This is important information. In addition, an activity analysis of only the costs of the overhead department of purchasing is given in Figure 8.2.

Fig. 8.2 Activity-based analysis of costs of purchasing department

	£000
Request competitive bids	120
Vet suppliers	70
Agree contracts	90
Place purchase orders	140
Liaise with suppliers	80
Resolve problems	150
Total	650

Just as thinking of the functions rather than the parts of a product, so thinking of the activities rather than the expense items of an overhead gives a very different perspective. For example, Figure 8.2 raises the question, why is £150 000 spent on resolving problems?

Another aspect of activity-based management is to classify the activities as value added or non-value added activities, i.e. which activities add value for the customer. Many activities might add value for the organization but not for the customer. For example, in Figure 8.2 the activity in the purchasing department of 'resolve problems' costing £150 000 does not add any value for the customers because the problems should not have arisen in the first place. After classifying activities into value added and non-value added categories, the aim then is to reduce the amount spent on such non-value added activities or even to eliminate such activities.

SVA OF OVERHEADS – PURCHASING

Strategic value analysis of overheads uses elements of both zero-base budgeting and activity-based management. For example, as in zero-base budgeting, SVA starts from scratch by considering the functions delivered by an overhead area. Similarly, as in activity-based management, SVA considers the value of the functions to customers. However, how do you conduct a SVA of overheads?

As with all SVA exercises this is a team effort. One difference with overheads is that at least one team member will be a representative of customers because the overhead is simply a service provided to *internal* customers. So, for example, manufacturing managers are customers of various overhead services such as administration, maintenance and personnel. The objective of SVA in relation to overheads is not only to reduce the overhead cost, but also to improve the overhead service provided to other departments in the organization. One such overhead is purchasing where the team members could include, for example, an engineering manager, management accountant, manufacturing manager, purchasing manager and sales manager.

When an overhead area such as purchasing becomes very complex or expensive, then SVA can be a useful approach. As usual, having selected the overhead area for SVA, the second step is to set an objective such as maintain the quality of the purchasing overhead service but reduce the cost of providing that service by 'one-third, i.e. from £3 million to £2 million'. The specific purchasing service in this example is the purchase of direct materials which involves departments such as purchasing, production planning, stores and accounting (see Yoshikawa et al., 1994).

The third step is to collect information about the purchasing of direct materials such as the cost of this overhead service, which is £3 million per year, and details of the services provided by purchasing of direct materials. The fourth step is to decide the functions of the overhead service of purchasing. As usual, the aim is to express the functions in terms of a verb and a noun such as:

- provide materials
- meet production schedule
- control costs
- assure quality.

These functions are then linked together in a functional family tree for the purchasing of direct materials (see Figure 8.3).

The check on the logic of this functional family tree is to ask 'how', going from left to right in Figure 8.3, and 'why', moving from right to left in this functional family tree. For example, how do we provide direct materials? The answer is by meeting schedule. How do we meet the schedule? The answer is by providing correct volume of materials and by providing these materials on time. Similarly, going from right to left back through the functional family tree in Figure 8.3, why do we provide the materials on time? The answer is to meet the schedule. Why do we meet the schedule? The answer is to provide materials.

The fifth step is to calculate the cost of each function. One way to cost the functions of an overhead service such as purchasing is to cost the various activities involved, as in activity-based costing. For example, purchasing would have the activities listed in Figure 8.4.

The existing cost of the purchasing functions are shown in Figure 8.5 where the cost of the third level functions are not shown to keep the figures relatively simple.

Fig. 8.3 Purchasing functional family tree

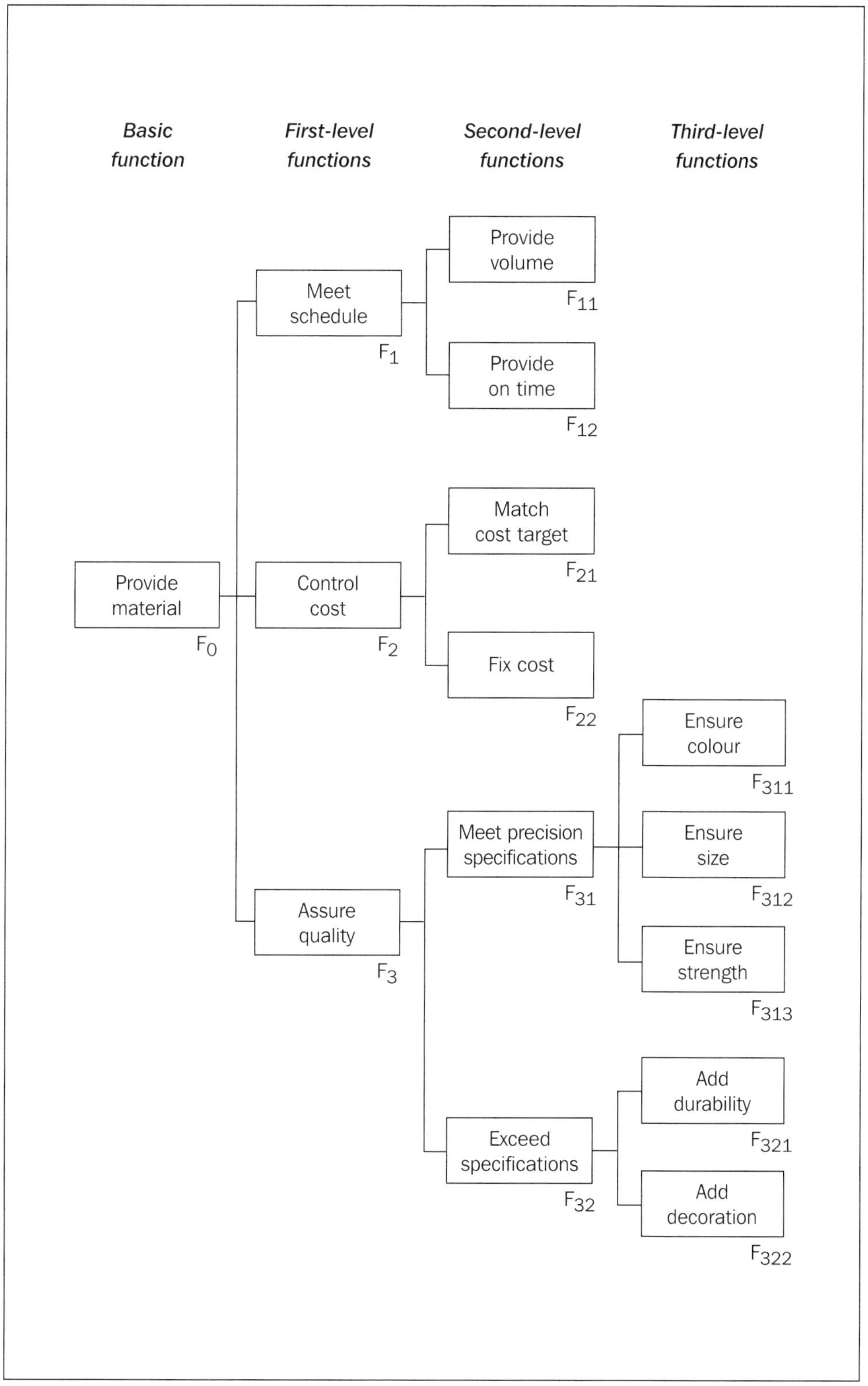

Basic function | *First-level functions* | *Second-level functions* | *Third-level functions*

Provide material F_0

Meet schedule F_1

Provide volume F_{11}

Provide on time F_{12}

Control cost F_2

Match cost target F_{21}

Fix cost F_{22}

Assure quality F_3

Meet precision specifications F_{31}

Ensure colour F_{311}

Ensure size F_{312}

Ensure strength F_{313}

Exceed specifications F_{32}

Add durability F_{321}

Add decoration F_{322}

Source: Yoshikawa et al. (1994)

Fig. 8.4 Activities in purchasing

- Assess material requirements.
- Search for suppliers.
- Vet potential suppliers.
- Gather price data.
- Negotiate prices with existing suppliers.
- Set and monitor budgets.
- Requisition and order materials.
- Receive materials.
- Inspect materials.
- Return materials.
- Move materials.
- Control stock of materials.
- Deliver materials to production line.
- Expedite suppliers – quality.
- Expedite suppliers – delivery.
- Pay suppliers.
- Maintain documentation.
- Manage overall process.

Fig. 8.5 Existing cost of purchasing functions

First-level functions			Second-level functions		
		£000			£000
F_1	Meet schedule	1080	F_{11}	Provide volume	410
			F_{12}	Provide on time	670
					1080
F_2	Control cost	730	F_{21}	Match cost target	580
			F_{22}	Fix cost	150
					730
F_3	Assure quality	1190	F_{31}	Meet specifications	1000
			F_{32}	Exceed specifications	190
					1190
		3000			

Figure 8.5 shows that the existing cost of purchasing direct materials is £3 million. The sixth step is to determine the customers' values for each function. This is easier for overhead services because the customers are within the organization, namely, managers using the service provided by purchasing. In this example, the managers rated the second-level functions out of 100 per cent as in Figure 8.6.

Fig. 8.6 Customers' views on purchasing functions

	%
Provide volume	20
Provide on time	35
Match cost target	10
Fix cost	0
Meet specifications	35
Exceed specifications	0
Total	100

Figure 8.6 reveals that the users of the purchasing service did not value the functions of fix cost or exceed specifications.

Having determined the customers' values of the functions, the target cost of £2 million for the purchasing of direct materials can be assigned to each function using these customer derived percentages. The assigned target cost and the existing costs for the second-level functions are shown in Figure 8.7.

Fig. 8.7 Existing and target costs for purchasing functions

Existing cost		Assigned target cost	
	£000		£000
Provide volume	410	20% × £2 million =	400
Provide on time	670	35% × £2 million =	700
Match cost target	580	10% × £2 million =	200
Fix cost	150	0% × £2 million =	0
Meet specifications	1000	35% × £2 million =	700
Exceed specifications	190	0% × £2 million =	0
	3000		2000

The problem functions can be determined from Figure 8.7 during the eighth step where the existing cost of providing that function is much greater than the customer assigned target cost namely:

- match cost target
- fix cost
- meet specifications
- exceed specifications.

For the above four functions the existing cost is much greater than its value to the customer (i.e. the assigned target cost). Again it is important to have this customer perspective on the overhead service.

The ninth step is the brainstorming session. In this example of purchasing direct materials, the chosen solution was to select fewer suppliers and involve the chosen suppliers much more closely with the company. For example, a new activity was added of running joint courses with some suppliers in total quality control (TQC) and also SVA. In addition, the SVA team recommended that various activities should be eliminated, such as inspection, storage and handling. This was because more reliance would be placed on the suppliers' own quality control and suppliers also would deliver directly to the production line on a just-in-time basis and in such a way that the materials would be used immediately. Of course, the suppliers would incur some extra costs but these were more than offset by the larger and longer-term contracts awarded to them.

The new simplified functional family tree is shown in Figure 8.8. The reduced number of activities together with more explanations of the changes are given in Figure 9.9.

Fig. 8.8 New simplified purchasing functional family tree

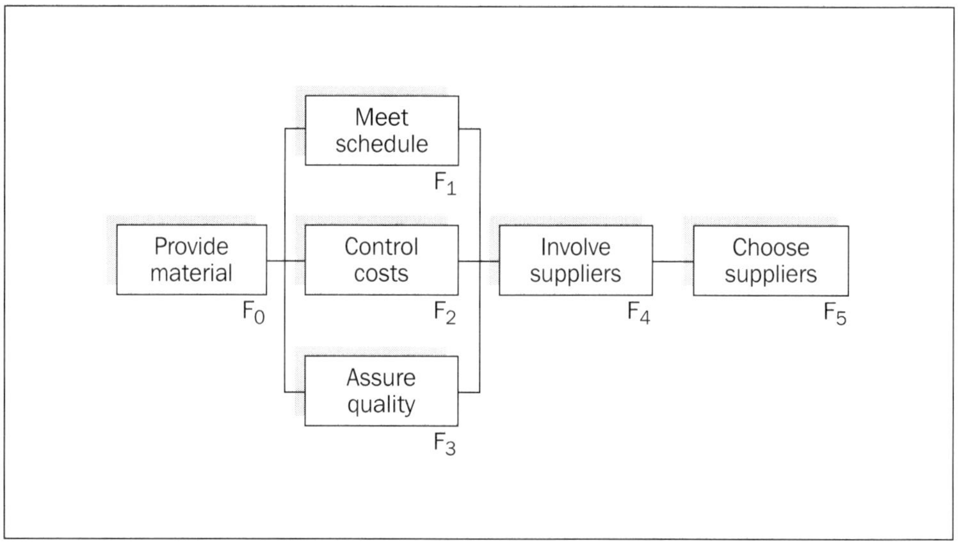

Source: Yoshikawa et al. (1994)

Fig. 8.9 Activities in new simplified purchasing family tree

Activities	*Comments*
Assess material requirements	
Negotiate price – existing suppliers	Easier because more mutual understanding
Set and monitor budgets	More certainty with longer-term supplier relationships
Requisition and order materials	More frequent ordering but less bureaucratic
Receive materials	More frequent deliveries but direct to production line
Pay suppliers	
Maintain documentation	Simplification
Manage overall process	Simplification
Run joint courses with suppliers in TQC and SVA	New activity
Search for new suppliers	Reduced activity due to longer-term relationship with existing suppliers
Vet potential suppliers	Reduced activity as above
Negotiate prices with potential suppliers	Reduced activity as above

Comparing the activities in Figure 8.9 with the activities in Figure 8.4 shows that the SVA exercise has reduced the number of activities from 19 to 12 and a number of the remaining 12 activities have been simplified.

The annual cost of the purchasing of direct materials with the new, simpler functional family tree is shown in Figure 8.10.

Fig. 8.10 Revised cost of purchasing after SVA

	£000
Meet schedule	1000
Control costs	150
Meet specifications	500
Total	1650

Figure 8.10 shows that the target cost of £2 million was more than achieved with the purchasing of direct materials now costing £1 650 000 instead of £3 million. In addition, the quality of this purchasing service has improved with:

- closer co-operation with suppliers
- longer-term relationships with suppliers
- simplified documentation
- simplified purchasing process
- material deliveries direct to production line
- joint courses in TQC and SVA run with suppliers.

Strategic value analysis is a particularly helpful approach not only in reducing overhead costs but also in developing new and better ways of providing an overhead service. In particular the views of the customers of that overhead service are given full consideration in the redesign of the way in which that service is provided. Strategic value analysis is a useful approach to the management of overheads.

SVA OF SERVICES

Just as SVA can be applied to products and overhead services, so SVA is a useful approach for service organizations such as banks, educational institutions, government organizations, hotels, insurance companies and transport organizations. As with overhead services, the only difference from applying SVA to services as distinct from products is that activities are equivalent to the physical parts of products. The detailed example of applying SVA to the overhead service of purchasing has already been discussed and the same process applies to external services as to internal overhead services.

An example of applying SVA to services, is given in Figure 8.11 for a restaurant (i.e. excluding the kitchen activities).

Figure 8.11 shows the sixteen activities involved such as welcoming the customer to the restaurant, showing the customer to the table, giving the customer a menu, etc. through to taking payment for the meal, thanking the customer, asking the customer to visit again, clearing the table and setting the table again. After an SVA exercise, the number of activities was dramatically reduced by, in effect, turning the restaurant into a fast-food outlet. Figure 8.12 lists the revised four activities following this SVA exercise.

Fig. 8.11 List of restaurant activities (excluding kitchen)

1. Welcome customer.
2. Show customer to table.
3. Give customer menu.
4. Take drinks order.
5. Serve customer bread rolls and water.
6. Take food and wine order.
7. Serve wine.
8. Serve food.
9. Clear dishes.
10. Ask for customer's views on meal.
11. Give bill.
12. Take payment.
13. Thank customer.
14. Ask customer to visit again.
15. Clear table.
16. Set table.

Fig. 8.12 Revised restaurant activities after SVA exercise

1. Take food and drink order.
2. Serve food and drink.
3. Take payment.
4. Clear tables.

9

Performance measurement and decision-making

PERFORMANCE MEASUREMENT

The framework of the balanced scorecard

The business environment has changed greatly during the 1990s. In some markets demand is lower than supply so that prices have fallen (such as television sets). A second change is that customers are not satisfied with standardized products but, instead, they wish for customized products.

A third change is that national boundaries are no longer obstacles to competition, so global management is essential. A fourth change is that product life cycles are getting shorter and shorter, so innovation and speed to market are the keys to success – an example is the personal computer. A fifth crucial change is that the future is not simply an extension of the past, so it is very difficult to manage organizations. Strategic management becomes very important. Many strategic management systems have been developed and marketed, including the balanced scorecard.

The balanced scorecard was developed by Robert Kaplan and David Norton, and their first paper was published in the *Harvard Business Review* in 1992. Kaplan and Norton (1992, p.72) suggested, 'think of the balanced scorecard as the dials and indicators in an airplane cockpit ... Similarly the complexity of managing an organization today requires that managers be able to view performance in several areas simultaneously'.

Kaplan and Norton also published their book, *The Balanced Scorecard*, in 1996, in which they argue that the balanced scorecard 'retains an emphasis on achieving financial objectives but also includes the performance drivers of these financial objectives. The scorecard measures organizational performance across four balanced perspectives: financial, customers, internal business processes, and learning and growth' (Kaplan and Norton, 1996, p.2).

Kaplan and Norton (1996, p.18) suggest that the balanced scorecard 'is a new framework for integrating measures derived from strategy'. However, the balanced scorecard is more than a measurement system – it is also a management system. The balanced scorecard is a strategic management system to enhance competitive advantage and profitability by breaking down vision and strategy into specific actions. The balanced scorecard tries to communicate vision and strategy not only for top management, but also for everybody in the organization. The balanced scorecard clarifies and adds consensus about vision and strategies.

One aspect of the balanced scorecard is the financial perspective. In order to be successful financially, organizations need a customer perspective. To improve customer satisfaction, organizations have to have excellent internal business processes such as an innovation process (developing products or services and designing them based on customer needs), operations process (manufacture products or services), marketing process and after-sales service process. Therefore, they need an internal business process perspective. Finally, to excel at certain business processes, organizations need innovating and learning capabilities. Therefore, they need a learning and growth perspective.

Top management of companies can plan an organization's future, identify problems and take corrective action by performing speedy and systematic measurements, analysis and reviews from the four perspectives of the balanced scorecard. The framework of the balanced scorecard is shown in Figure 9.1 (see Bullen and Rockart, 1981, figures 1 and 9 in relation to aspects of critical success factors).

Fig. 9.1 Framework of the balanced scorecard

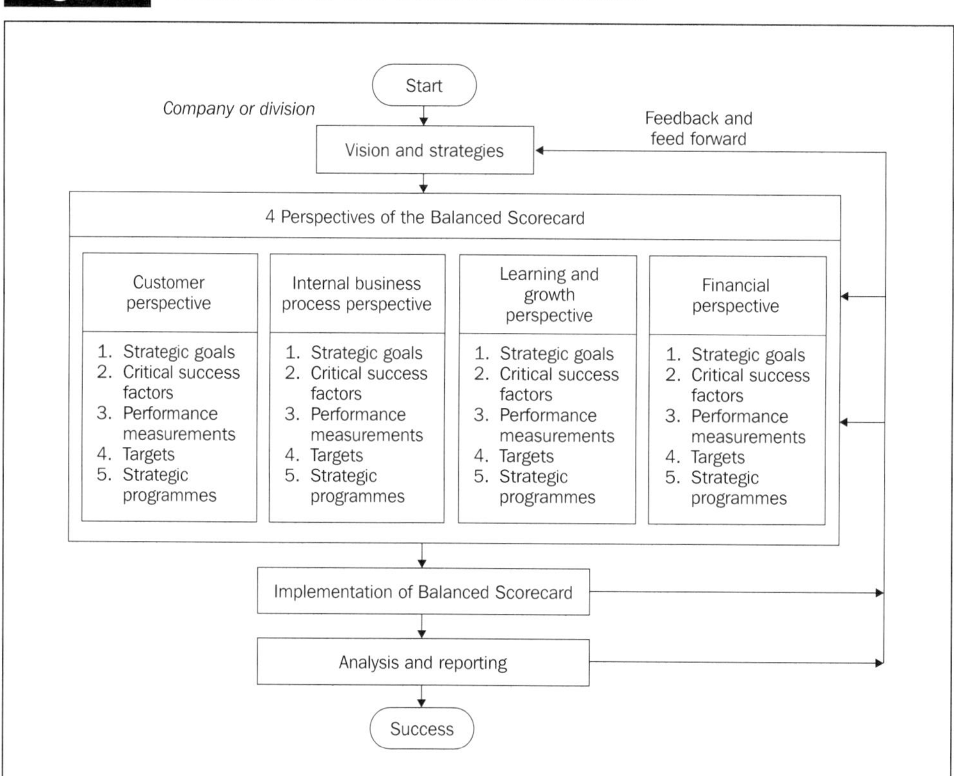

SVA and performance measurement

The balanced scorecard can be built by the following seven steps (Olve et al., 1999, pp. 49–77).

Step 1 is to establish the vision and strategies of the organization.

Step 2 is to establish the perspectives by critical success factor analysis.

Step 3 is to break down the vision and strategies into each perspective and formulate overall strategic goals and maps.

Step 4 is to identify critical success factors from strategic goals.

Step 5 is to develop performance measurements and identify causes and effects.

Step 6 is to establish the target for each performance measurement.

Step 7 is to breakdown the corporate level balanced scorecard into the appropriate level of organizational unit such as division, department and section.

Once the balanced scorecard is built, it is implemented and the performance is analysed and reported.

A crucial step is step 3 above, which is to break down the vision and strategies into each perspective and formulate overall strategic goals and maps. Strategic value analysis can help in this process, as will be illustrated by the example of the Southwest Airlines balanced scorecard from Krieger and Gregory (2001) (see Figure 9.2).

Fig. 9.2 Southwest Airlines balanced scorecard

Strategic theme: operating efficiency	Objectives	Measurement
Financial: Profitability; Lower costs; Increase revenue	■ Profitability ■ Increase revenue ■ Lower costs	■ Market value ■ Seat revenue ■ Plane lease cost
Customers: Flight is on time; Lowest prices	■ Flight is on time ■ Lowest prices	■ FAA on time arrival rating ■ Customer ranking (market survey)
Internal: Fast ground turnaround	■ Fast ground turnaround	■ On ground time ■ On time departure
Learning: Ground crew alignment	■ Ground crew alignment	■ % ground crew as shareholders ■ % ground crew trained

Source: Krieger and Gregory (2001)

The strategic theme of Southwest Airlines is operating efficiently to increase profitability. One question is how do they increase profitability from a financial perspective point of view. They may find a variety of ways to do it by using SVA. One example is shown in Figure 9.3. There are two major ways to do it. They can increase profitability by increasing revenue and/or lowering costs.

BISHOP BURTON COLLEGE
LIBRARY

Fig. 9.3 Strategic goals from financial perspective

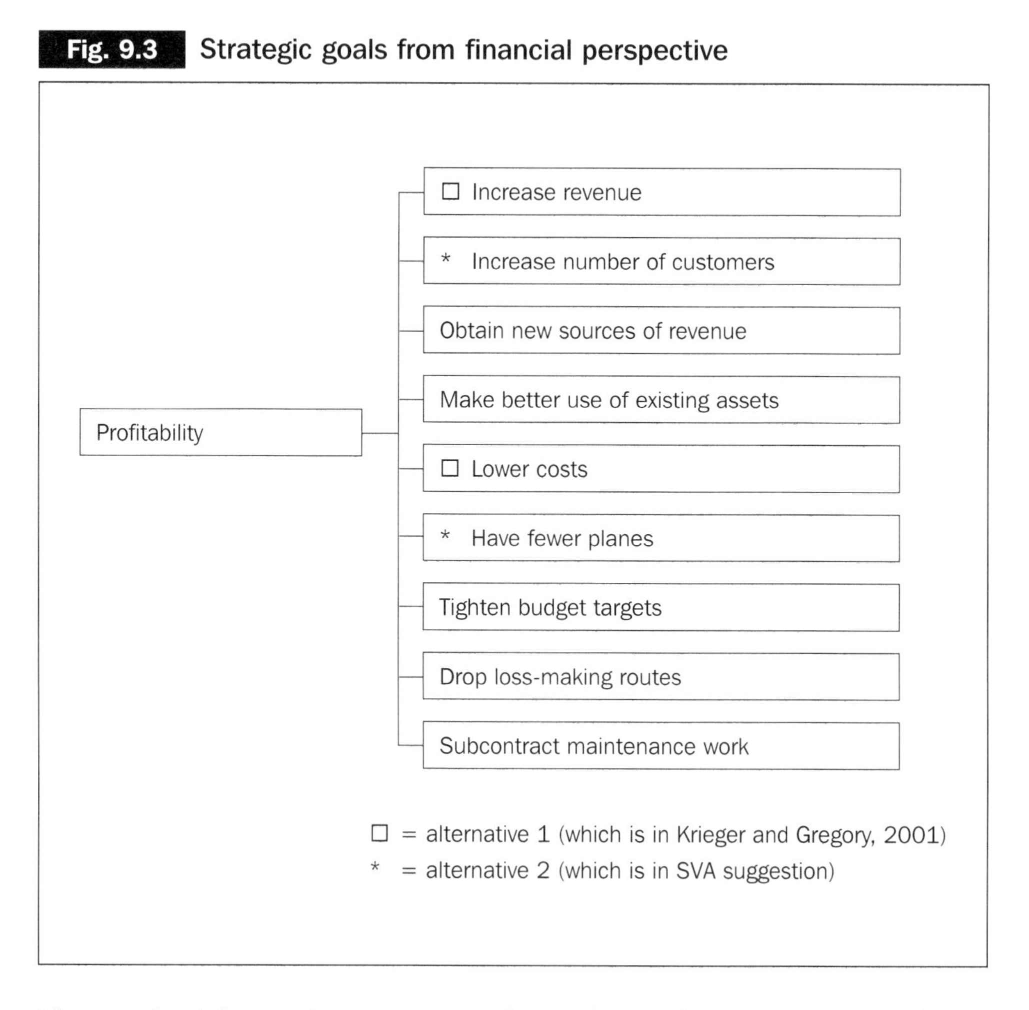

The question is how to increase revenue from a financial perspective point of view. Alternatives include increase the number of customers, obtain new sources of revenue and utilize existing assets more efficiently. The ways to lower costs include alternatives such as operate with fewer planes, tighten budget targets, drop loss-making routes and subcontract maintenance work (see also Bullen and Rockart, 1981, figures 1 and 9).

If they choose two alternatives such as increase revenue and lower costs from a financial perspective, they have to find ways to do that from a customer perspective point of view. According to Figure 9.4, they can increase their revenue in many ways such as have lowest prices, increase customer loyalty, create brand image, speed up check-in, tender for in-flight catering supplies, improve service, extend electronic check-in, minimize baggage loss, give higher mileage points, offer comfortable seats, have new advertising campaign, make attractive package tours, set up deals linked to other products, give friendlier service and ensure flight is on time. They can also lower costs from a customer perspective as shown in Figure 9.5 by ensuring flight is on time, reducing staff turnover, introducing Internet booking, automating administration and check-in by local automatic ticket machine.

Fig. 9.4 Strategic goals from customer perspective (revenues)

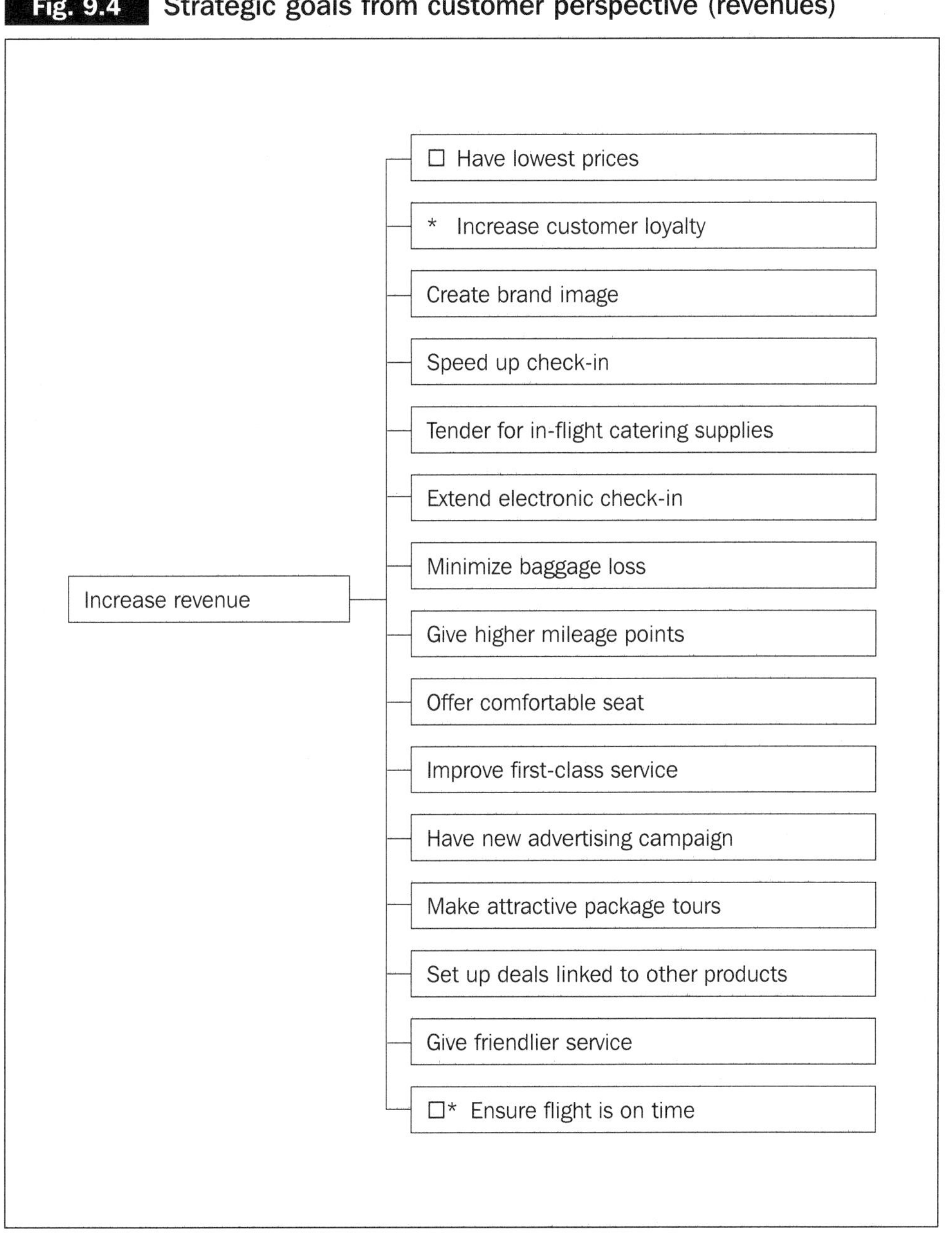

The next question is how to have lowest prices from an internal perspective point of view. According to Figure 9.6, there are a variety of alternatives including faster ground turnaround, keep to schedule, have joint flights, reduce overhead, take fuel-economy measures, reduce office space, increase capacity usage, benchmark lowest price competitor and charter aircraft for peak periods.

Figure 9.7 shows alternatives of ensuring flights are on time from an internal business perspective and Figure 9.8 illustrates alternatives of faster ground turnaround from a learning and growth perspective.

Fig. 9.5 Strategic goals from customer perspective (costs)

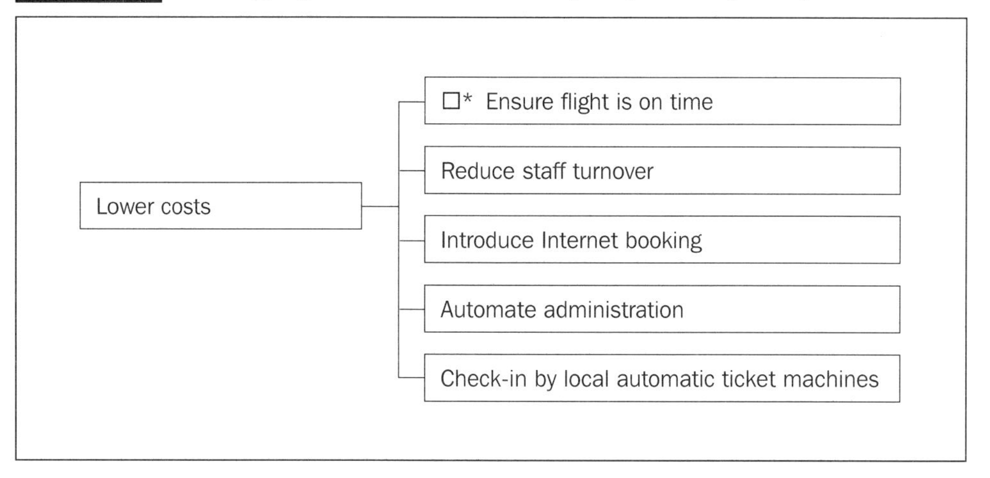

Fig. 9.6 Strategic goals from internal business process perspective (prices)

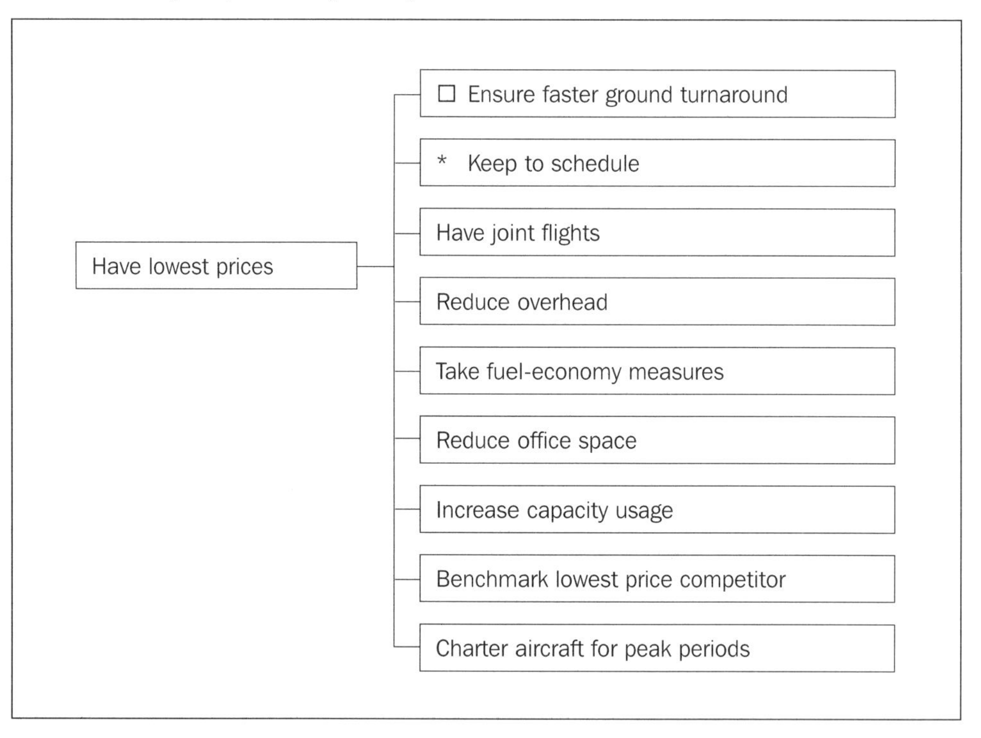

In the case of Southwest Airlines, it could draw a strategic map such as alternative 1 (Krieger and Gregory, 2001) in Figure 9.9. However, an advantage of the SVA approach is that it can show many alternatives for a strategic map. One of the alternatives is alternative 2 in Figure 9.9 based on increasing the number of customers while at the same time having fewer planes. It is important that the alternative chosen is based on the organization's own vision and strategy. Strategic value analysis can assist in this process by clarifying the possible range of alternatives.

Fig. 9.7 Strategic goals from internal business process perspective (on time)

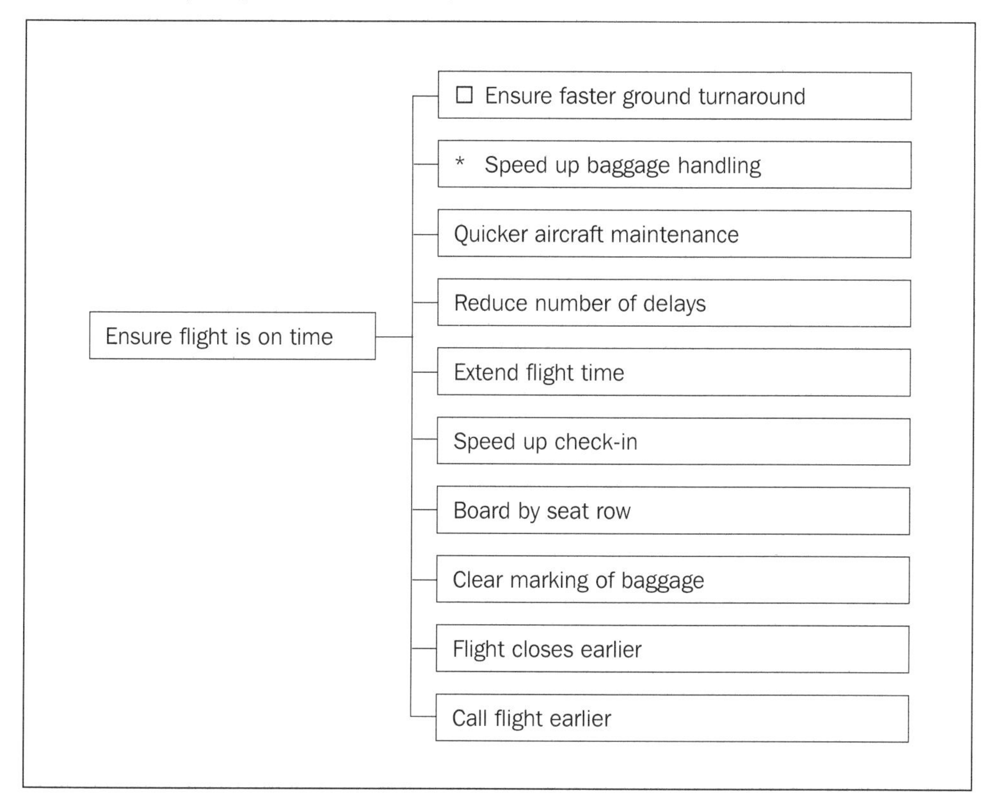

Fig. 9.8 Strategic goals from learning and growth perspective

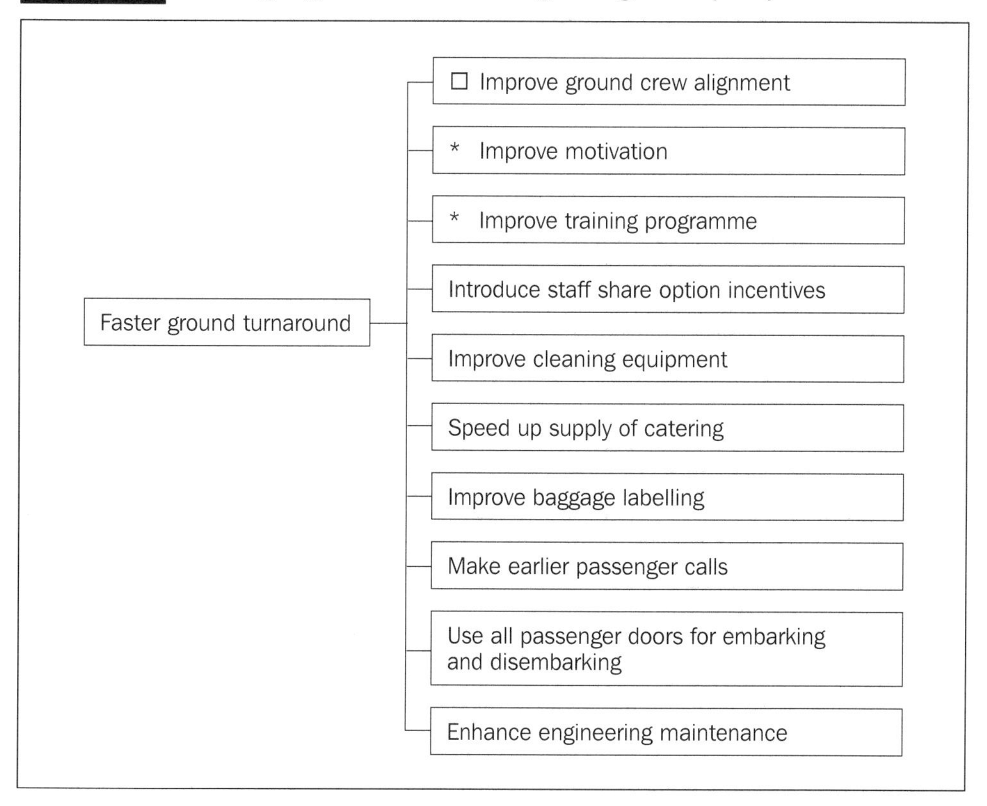

Fig. 9.9 Strategic maps

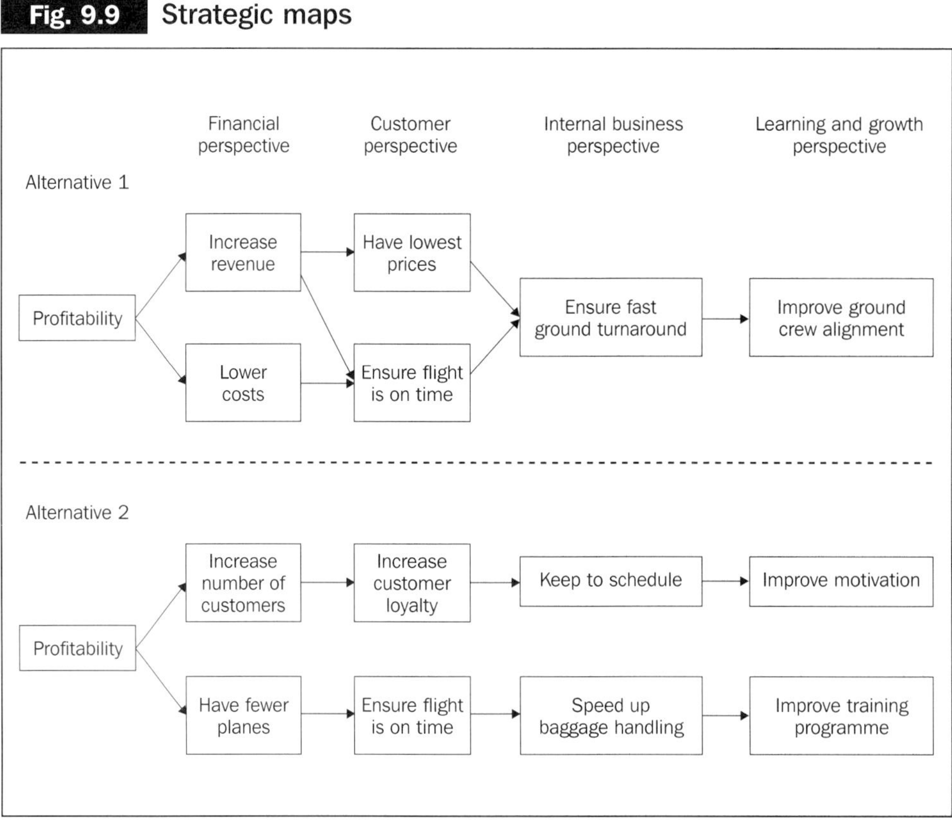

Once a strategic map has been drawn, the next step is to set up performance measurements for each strategic objective. The Southwest Airlines strategic map and performance measurements are shown in Figure 9.2. An alternative strategic map and performance measurements are illustrated in Figure 9.10 using SVA.

DECISION-MAKING

As can be seen from the above, SVA can provide strong support for the decisions required to support an organization's strategic mapping activity. However, it can also provide a more direct framework for all types of decision analysis through the construction of decision-based functional family trees based on means–end relationships. This type of decision support can support either the identification of decision alternatives or the economic measurement of these alternatives.

Identification of decision alternatives

Decision analysis can be based on the construction of a decision tree which clearly presents the variety of future action paths that stem from any decision point. This approach not only forces the decision maker to consider the identification of alternative possibilities (a key characteristic of decision-making rationality) but also highlights the implications which flow from embarking on a particular decision path. The SVA technique can be instrumental in constructing the decision tree which, in essence, is similar to the functional family tree diagrams which are central to SVA. The process is explained below using the decision to make or buy in a new component.

Fig. 9.10 **Strategic map, strategic objectives and performance measurements**

Strategic theme: operating efficiency	**Strategic objectives**	**Performance measurements**
Financial perspective Profitability Increase number of customers → Profitability Have fewer planes → Profitability	■ Profitability ■ Increase number of customers ■ Have fewer planes	■ ROE ■ Net profit ■ Growth of sales ■ Number of new customers ■ Number of flights ■ Number of planes ■ Leasing costs
Customer perspective Increase customer loyalty → Increase number of customers Ensure flight is on time → Have fewer planes	■ Increase customer loyalty ■ Ensure flight is on time	■ % of repeat customers ■ Growth of customers ■ % of flights on time ■ Average period of lateness
Internal business perspective Keep to schedule → Increase customer loyalty Speed up baggage handling → Ensure flight is on time	■ Keep to schedule ■ Speed up baggage handling	■ % of flight cancellations ■ Customer complaints ■ Average time taken in loading ■ % of times loading meets target time
Learning growth perspective Improve motivation → Keep to schedule Improve training programme → Speed up baggage handling	■ Improve motivation ■ Improve training programme	■ % staff in bonus scheme ■ Staff survey results ■ Training cost as % of sales ■ Training personnel hours

Step 1

Decision involves the securing of a source of supply for a new component X. This involves two options namely (1) make it in house or (2) purchase it from a supplier. These are identified by asking the question 'how?' at the original decision point in order to generate the options.

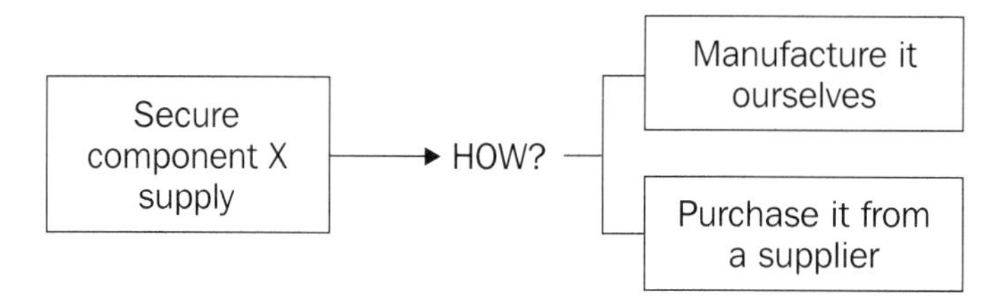

It should be noted that the selection of an original decision point does constrain the scope of the decision. In the above example it assumes securing a component X supply as a given, and options such as redesigning the final product to exclude it are not encompassed in the resulting decision analysis. Thus the selection of an initial starting point for the analysis is an extremely important part of the process.

Step 2

(a) For each of the initial two decision options further options for their possible achievement can be derived by again questioning 'how?'

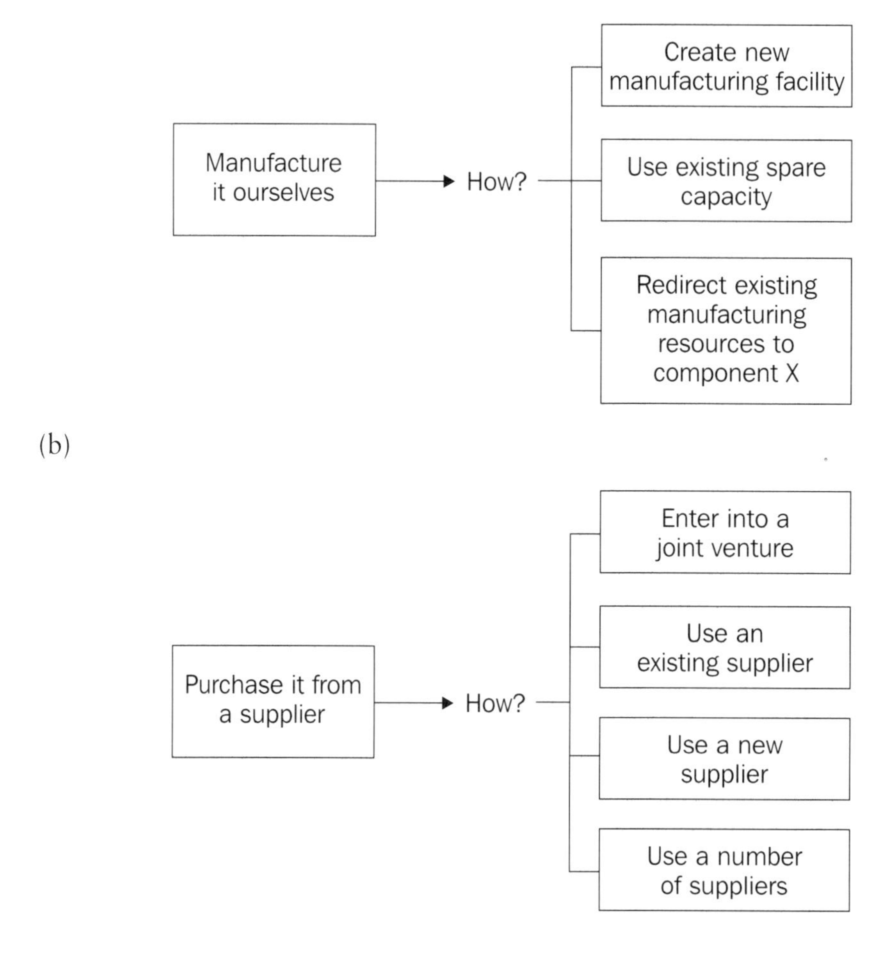

Further steps can be added while the question 'how?' continues to generate meaningful options. Where this no longer occurs then the action has become so specific that it does not allow for any optional ways of achieving it.

As one moves from left to right in the decision tree, the decision tends to change in nature from being significant strategically and pertinent to higher-level management, to one which eventually becomes more routine or operational. At the far right of the final decision tree are a list of the practical possibilities facing the decision maker (see the options in Figure 9.11).

Fig. 9.11 **Final decision tree**

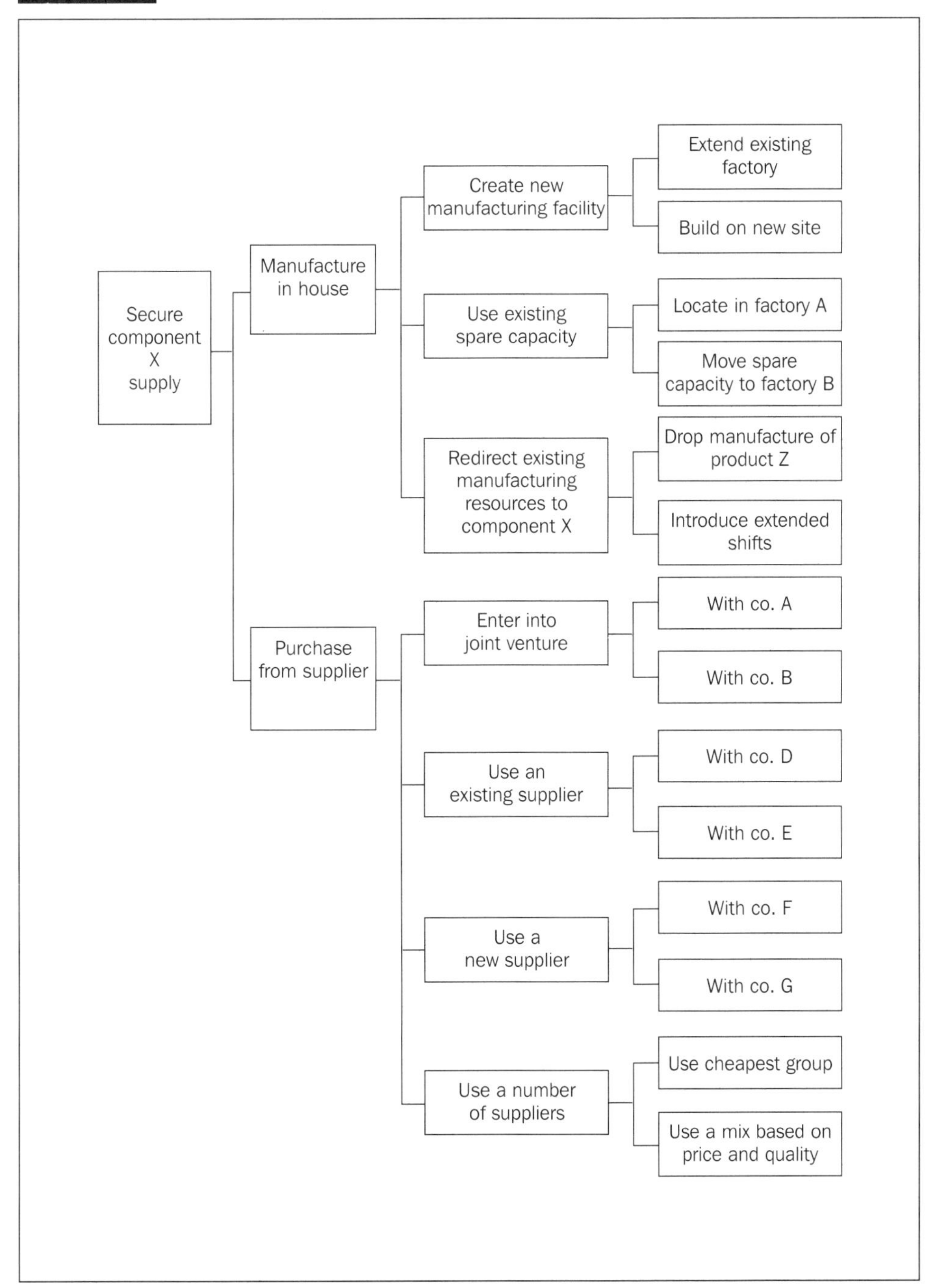

Step 3

Obtain relevant decision data for each of the decision possibilities. These should include:

- strategic fit
- viability
- timescale for achievement
- financial impact.

From consideration of the options and their characteristics a preferred selection may be obtained. The process of identifying and assessing the possibilities is one which also adds to an appreciation and understanding of the organization and its situation. It can stimulate ideas for change and innovation at various decision levels as well as requiring decision makers to become familiar with strategy, organizational needs and financial structures. The decision structure has to be constructed and the final selection of an option reveals the 'pathway' which highlights all of the key implications of making the decision.

Financial impact

Decisions involve sacrifices and it is these sacrifices which give rise to the costs which constitute a major part of the decision's financial impact. Where a decision (as is commonly the case) involves a commitment (and consumption) of resources then this represents the sacrifice or relevant cost. Quantifying this sacrifice does, however, require some analysis and it is here that the SVA approach can play a part.

The sacrifice of a resource can be measured by the use of the concept of deprival value. This concept was originally developed in a legal context in order to obtain loss measurement for compensation purposes. It is based on the notion that an asset (or resource) may have different values (or costs) depending on the circumstances pertaining when it was lost (or consumed). This approach to cost ascertainment can be represented in a functional family tree and can be developed step by step to show the decision relevant cost.

Step 1

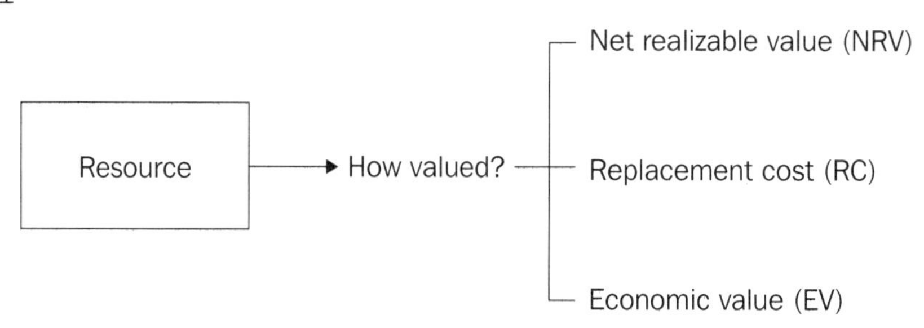

Three possible values exist for any resource and in particular circumstances any of them could represent the cost for decision purposes. The relevant cost is determined by ascertaining the loss suffered by the decision maker (or their organization) when they are deprived of the resource (through its decided use). If they can replace the resource then the maximum possible loss is its replacement cost (RC) as they can always reinstate the resource for this price. Thus where EV and/or NRV exceed RC then RC is the relevant decision cost. However, it may be that the resource does not merit replacement. This occurs where the resource's sales value (NRV) and its value in its best alternative economic use (EV) are less than the replacement cost. In these circumstances the resource has a worth (and should be valued for decision purposes) which is the higher of NRV and EV (which represent its best value to the organization).

To complete the determination of the cost impact of the decision some consideration has to be given to the identification of these three values. This may be done by the incorporation of a second step or indeed further extensions to the functional family tree (see Figure 9.12).

Fig. 9.12 **Cost determination: functional family tree**

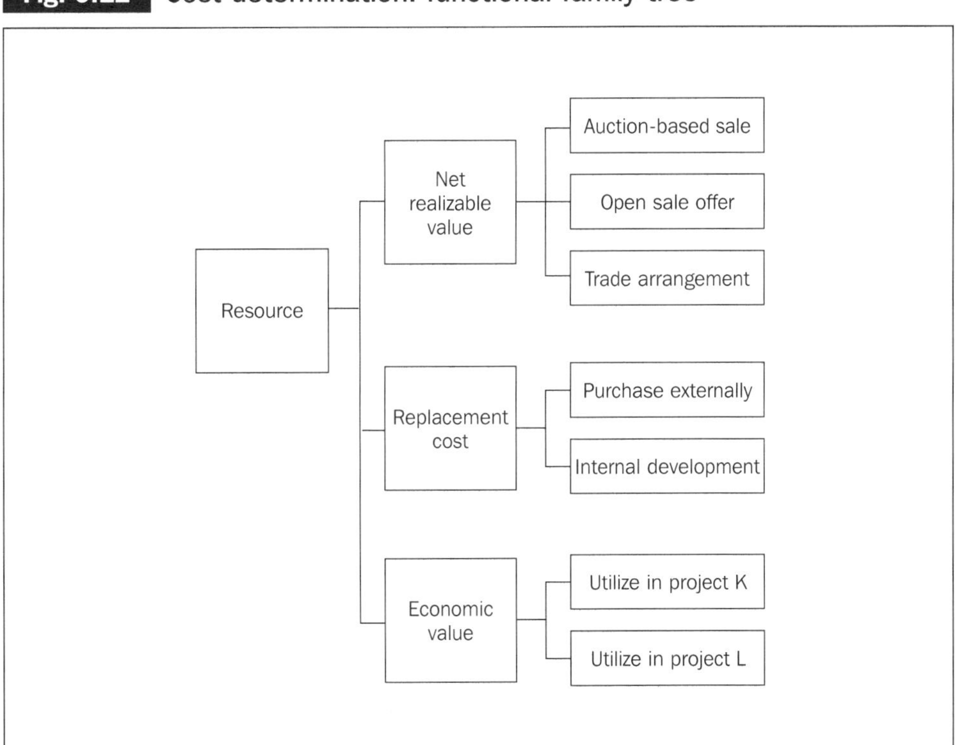

This presentation requires the decision maker to approach the decision from an opportunity cost perspective. This is the most rational economic basis for taking decisions. It ensures a range of options are identified. The two common options

of replacement or sale can be determined readily where wholesale and retai
markets exist for the resource. The economic value is more subjective both
terms of identifying the alternative uses of the resource and then in quantif
their value. The latter task requires the discounting of the future cash fl
estimates from internal use of the resource at an appropriate cost of capital.

10

Conclusions

Strategic value analysis is a technique with multiple applications within the firm. It can assist in cost reduction for both direct costs and overheads, it can help in identifying areas for investment and the improvement of product or service functionality and it can contribute to the derivation of strategic plans and the decision analyses which underlie them. In all these applications it provides the basis for a systematic approach to achieving the relevant aims. It is thus essentially a practical and versatile technique.

As SVA is operated with multidisciplinary participants, it provides a means of integrating the different specialisms in pursuit of specific purposes. It thus harnesses the intellectual capital of the firm to produce the interactive synergies which will deliver creativeness and innovation. Moreover, this is achieved in a context of awareness of the external constraints and requirements of the market in which the firm operates. It thus ensures that a strategic element is impounded in the actions taken as a result of SVA activities.

Strategic value analysis can be applied to products and services. It is a useful approach both for existing and for new products and services. The essence of SVA is that it concentrates on the functions rather than on the parts of a product or service. Another essential feature of SVA is that it is a team approach with a very structured approach, including a set of detailed steps and a number of detailed worksheets. However, it is important to remember that the important output from this structured approach is the results of the brainstorming session as illustrated by the case studies in this book.

In addition to external services, SVA is an approach which can help with the management of internal services, namely, overheads. Again an important aspect of SVA is that it incorporates customers' views as central to the whole process. Strategic value analysis is a wide-ranging approach which uses not only costs but also customers' values.

Finally SVA complements many of the recent new techniques developed in accounting and management. It facilitates the development of quality initiatives, increases flexibility and speed in working practices, improves customer orientation, utilizes activity-based costing and management information, and provides a means of devising the performance measurement structures underlying the balanced scorecard. It is thus an analytic technique which adds value to decision-making and which helps to promote an organization's strategy – hence the name, strategic value analysis.

Bibliography

Bullen, C. V. and Rockart, J. F. (1981) *A Primer on Critical Success Factors*, CISR, no. 69, Massachusetts Institute of Technology.

Creasy, R. (1973) *Functional Analysis System Technique Manual*. Irving: Society of American Value Engineers.

Defence Equipment Society (DES) (1989) *Procurement Manual of Defence Agency Central Procurement Office*. Tokyo: Defence Procurement Studies Committee, 159–246 (in Japanese language).

Government Buildings Department of Minister's Secretariat in Ministry of Construction (GBD) (1991) *Cost Estimation Handbook for Building Preventative Maintenance in Industry*. Tokyo: Building Preventative Maintenance Centre (in Japanese language).

Kaplan, R. S. and Norton, D. P. (1992) 'The balanced scorecard – measures that drive performance', *Harvard Business Review*, January–February, 71–9.

Kaplan, R. S. and Norton, D. P. (1996) *The Balanced Scorecard*. Boston: Harvard Business School Press.

Krieger, J. A. and Gregory, B. (2001) 'The balanced scorecard and six sigma – an integrated approach', presentation at Balanced Scorecard Collaborative Net Conference, February 2001, 9.

Mouritsen, A., Hansen, A. and Hansen, C. O. (2001) 'Episodes around target cost management/functional analysis and open book accounting', *Management Accounting Research*, June, 221–44.

Olve, N.-G., Roy, J. and Wetter, M. (1999) *Performance Drivers: A Practical Guide to Using the Balanced Scorecard*. Chichester: Wiley.

Phyrr, R. A. (1970) 'Zero-base budgeting', *Harvard Business Review*, November–December, 111–121.

Sanno Institute of Management VM Centre (1995) *Foundation of VE*. Tokyo: Sanno Institute of Management (in Japanese language).

Sato, R. (1965) *Cost Table*. Tokyo: Sangyou Nouritu Junior College (in Japanese language).

Yoshikawa, T. (1992) 'Comparative study of cost management in Japan and the UK', *Yokohama Business Review*, June, 79–106 (in Japanese language).

Yoshikawa, T., Innes, J. and Mitchell, F. (1989) 'Cost management through functional analysis', *Journal of Cost Management*, Spring, 14–19.

Yoshikawa, T., Innes, J. and Mitchell, F. (1990) 'Japanese cost tables', *Journal of Cost Management*, Fall, 30–6.

Yoshikawa, T., Innes, J. and Mitchell, F. (1994) 'Functional analysis of activity-based costing information', *Journal of Cost Management*, Spring, 40–8.

Yoshikawa, T., Innes, J. and Mitchell, F. (1995) 'A Japanese case study of functional cost analysis', *Management Accounting Research*, December, 415–32.

Yoshikawa, T., Innes, J. and Mitchell, F. (1996) 'Japanese cost management practices' in Brinker, B. J. (ed.) *Handbook of Cost Management*. Boston: Warren Gorham and Lamont, F3-1–F3-29.

Yoshikawa, T., Innes, J. and Mitchell, F. (1997) 'Performance measurement for cost management: the nature and role of Kousuu', *Journal of Management Accounting Japan*, 5, 2, 47–61.